> The fact that we live at the bottom of a deep gravity well, on the surface of a gas-covered planet going around a nuclear fireball 90 million miles away and think this to be normal is obviously some indication of how skewed our perspective tends to be.
>
> Douglas Adams, The Salmon of Doubt: Hitchhiking the Galaxy One Last Time

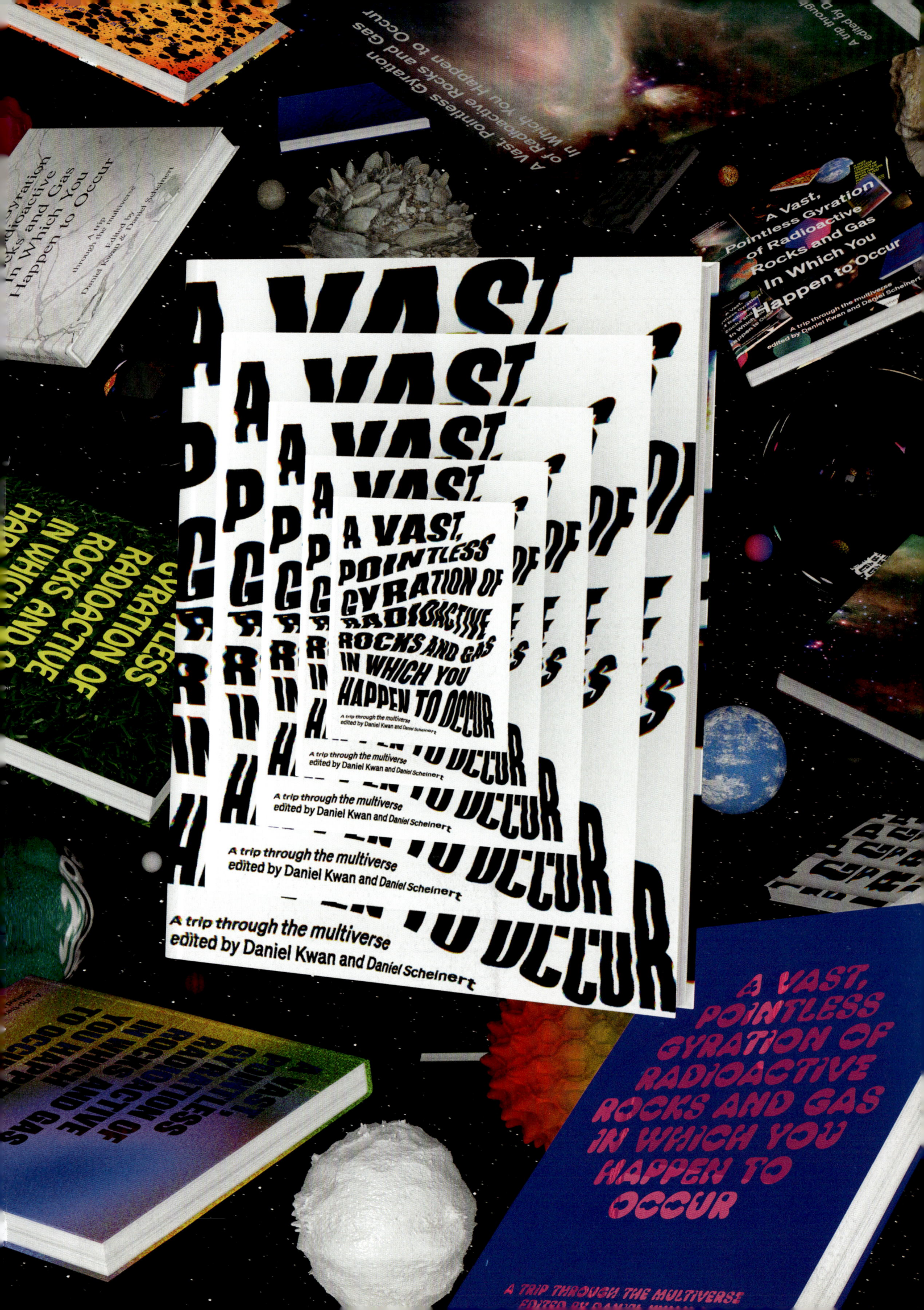

| 5 | **Foreword**
Ted Chiang

9 | **Introduction**
by Daniel Kwan and Daniel Scheinert

11 | *Everything Everywhere All At Once*
Alternate Take
by Daniel Kwan and Daniel Scheinert

32 | **Shock**
Esmé Weijun Wang

34 | **Broken**
Etgar Keret

36 | **Daybreak**
Emily Segal

38 | **27 Places**
Billy Chew, Julia Pott, and Daniel Kwan

52 | **The Many Worlds of Hugh Everett III**
Lizzy Stewart

68 | **The Patchwork Effect**
A roundtable of quilters discuss the quilted multiverse
Kelsey Keith with Becky Scheinert, Luann Johnson, and Rebecca Tait

95 | **Pop**
An essay on ego, the cosmos, and our next big demotion
Sasha Sagan

104 | **Daniel and Daniel with David**
An interview between two filmmakers and their favorite neuroscientist, David Eagleman
Daniel Scheinert and Daniel Kwan

134 | **Reversal**
David Eagleman

146 | **We As Organism**
Alan Watts

Illustration

42 | **Parallel**
Liam Cobb
Ilya Milstein
Jun Cen

76 | **Quilted**
Qieer Wang

84 | **Bubble**
Max Guther

113 | **Surface**
Jul Quanouai

136 | **Cyclical**
Derek Ercolano
Ram Han
Céline Ducrot
Robert Beatty
Jordan Moss

Foreword

by Ted Chiang

Why tell stories set in a multiverse? A single universe is huge by any reasonable standard, and surely we haven't exhausted all the possible stories that one universe can accommodate. What does a multiverse bring to storytelling?

For a lot of people, their first encounter with multiverse stories are superhero comics or movies. I think this is a reflection of the fact that, in a way, superhero stories are the modern version of myths and legends. People have told stories about King Arthur for centuries, and while you can ask which is the oldest one, it doesn't make sense to ask which is the definitive one; each generation has its own version. Something similar is true for superhero stories; every DC Comics fan has their own favorite origin story of how Superman and Lex Luthor first met. The multiverse gives writers a way to embrace this; it acknowledges that there are different versions of the story and incorporates them within the world of the story itself. It transforms a superhero from a character into an archetype, a pattern that transcends any specific telling.

Now we're seeing multiverse stories pop up that have nothing to do with superheroes or mythical figures. To my mind, this is the culmination of a long trend in storytelling: the growing recognition of the role of contingency in our lives.

For much of human history, stories reinforced the idea of fate. They told us that events unfolded the way they did because of destiny or because of the will of God. This was true not just of narratives presented as fiction, but also of narratives presented as history—battles were won because a nation was "destined" to be victorious. But as we entered the modern era, there was greater recognition of the importance of chance and individual agency in the course of events. This was reflected in fiction in many different ways, perhaps the most overt of which was the invention of alternate-history stories. What if the Confederacy had won the Civil War? What if the Roman Empire had never fallen?

Multiverse stories are sort of like alternate-history stories with the dial turned up to eleven. Instead of imagining one possible alternative, they imagine a thousand. And where alternate-history stories typically focus on military history, multiverse stories can focus on personal history. They're less concerned with Robert E. Lee at Antietam and more with an ordinary person simply going about their business. They illustrate how a person's life can be dramatically affected both by individual agency (what if they had accepted that new job offer instead of staying with the same company?) and by chance (what if they had missed the train that day?).

Being confronted with the role of contingency in our lives can be disconcerting, because there are aspects of the idea of destiny that even the most modern individuals find appealing. For example, it's more romantic to think that you were always fated to meet your one true soul mate than to think that you could have been happy with any one of a dozen people. It's more comforting to imagine that your parents were fated to conceive you than to accept that you are just one of the many possible children they could have had. Multiverse stories ask you to let go of these notions and many others, and relinquishing all of them can feel a bit like losing one's faith in God. It leaves some people wondering whether life has meaning at all.

On the other hand, multiverse stories can also dramatize the idea that, across the myriad lives you might have led, there will be certain commonalities. Perhaps there are character traits that recur in every version of you, attributes that can be said to define you no matter how your external circumstances change. Maybe it's not only superheroes who are archetypes; maybe there's an archetypical you as well, a pattern that transcends any specific path your life might have taken.

A basic principle of fiction writing is that character is revealed through action, and the same is true for real life: who you are as a person is defined by the choices you make. If there's a coherent "you" who extends across different branches of the multiverse, it's someone who tends to behave in a certain way when faced with a decision. And such decision points bring us back to the importance of contingency. Whether we're talking about global history or personal history, nothing was inevitable. There were countless moments when it could have gone a different way. Those were moments when character mattered, when the principles that people held guided their actions and had long-lasting repercussions.

Those pivotal moments don't exist only in your past; they exist in your future, too. Even now, there are many different paths your life could take, and while you don't have absolute control over what will happen next, neither are you a slave to fate. You can choose to stay the course, or you can choose an entirely new direction—whatever you do will have consequences for you and for those around you. Multiverse stories serve as a reminder that your decisions matter. Act accordingly.

Ted Chiang's first collection of short stories, *Stories of Your Life and Others*, has been translated into 21 languages, and the title story was the basis for the Oscar-nominated film *Arrival*, starring Amy Adams. His second collection, *Exhalation*, was chosen by *The New York Times* as one of "The 10 Best Books of 2019."

Introduction

by Daniel Kwan and Daniel Scheinert

Writing a story in the multiverse can be frustrating. For every decision your characters make, in the back of your mind (and in the back of your audience's mind), there exists a universe in which the opposite decision was made. This breaks every unspoken rule that makes a story compelling. No decisions have weight. Every consequence is watered down.

If we take Vonnegut's sage advice into consideration that "every character should want something, even if it is only a glass of water," imagine what happens to your story once the character finally gets that glass somewhere in the multiverse. When your character finally drinks from the glass, they are simultaneously dumping it on the ground. The glass breaks, the water turns to wine, and neither you nor your character knows what anyone wants anymore. They are lost in a sea of chaos, trying to project meaning onto the data points their poor, ill-equipped brains have collected, all while those data points leak through their fists like sand in a windstorm. In many ways, it's not too different from what it feels like to be alive today.

This is what we kept chasing after as we wrote the screenplay for our film *Everything Everywhere All At Once*—the infinite ways the multiverse felt relevant to our lives today. Could the multiverse be the right metaphor for the information overload and emotional fatigue we're all experiencing? How about a way for us to explore the Pandora's box that was the invention of the internet, and how it deepened the generation gap between us and our parents? Maybe it could be a useful way to talk about the ideological bubbles that have begun to form, each universe becoming a proxy for the contradictory belief systems that our society has created to cope with the noise? There is a reason why the multiverse has completely overtaken conversations in pop culture; from Marvel to *Fortnite*, the multiverse is everywhere, and yet there is still so much more to explore.

We believe that, in many ways, quantum mechanics is the closest thing we have today to the myths of our ancestors. Just as their stories sprung out of the unsolvable mysteries of lightning bolts observed in the sky or questions about what happened to the sun at night, we too, are touching the edges of humanity's collective understanding of how the world works, and then projecting ourselves onto the shrouded unknown.

Only an estimated five percent of the universe's matter is visible to us. The simple act of observing photons can change the fundamental way they behave. It might be statistically more likely that a brain with all of your memories accidentally formed in a random cloud of matter than for you to have actually existed. From where we stand, our science alone is not enough to comprehend what the fuck any of this means. This is where art, philosophy, and getting stoned at 2 a.m. with your friends comes in. This is what this book is here for. Not to give you a few quick science lessons on quantum physics or M-theory. It is here, in the absence of answers, to play with the possibilities.

We asked the many talented and inspiring contributors—scientists, poets, artists, illustrators, writers, and beyond—in this book to come play along. We are dancing in the void of the unknowable, hoping that something out there will take pity on us and grace us with some answers—be it a god or a new law of physics we have yet to name. Come dance with us.

```
<Cue the music>
```

Everything Everywhere All At Once: Intro (Alt) [Narrator]

By

Daniel Kwan & Daniel Scheinert

©2017

EXT. BLACK VOID

                    NARRATOR (V.O.)
          Here we are, in this moment, at the
          beginning.

FADE IN: We open on a screenplay page with the words you are
now reading typed upon them.

                    NARRATOR (V.O.)
          And because most beginnings are
          also often endings, it would be
          wrong for me not to point out that
          we are also here, at the end.

The NARRATOR'S VOICE flows with speed and precision, and the
pace of the edit follows suit.

We see a flurry of script pages on laptop screens, in
printer trays, deep inside wastebaskets. Many of them are
covered in notes. Perhaps you catch a glimpse of "Draft 1,"
"Draft 4," or "Draft 7 ver 3."

                    NARRATOR (V.O.)
          And because every moment would not
          be possible without the moment
          before it and is rendered
          unnecessary without the moment
          after it, we could say that the
          existence of everything that has
          happened and will happen hangs on
          the existence of this one moment.

We land back on a single page in a bound book.

                    NARRATOR (V.O.)
          This is it.
          This is everything.

A light buzzes. A dog barks. Some street noise.

                    NARRATOR (V.O.)
          Let us begin.

EXT. BARNHOUSE - DAY

TITLE: KING OF PRUSSIA, PENNSYLVANIA 1912, A UNIVERSE NOT
DISSIMILAR TO OUR OWN

W. T. WARREN (20s) puts on a prototype leather football
helmet. He is squared off against a farmhouse wall.

                                             (CONTINUED)

CONTINUED: 2.

                    NARRATOR (V.O.)
          Outside of a run-down barn, in the
          middle of King of Prussia,
          Pennsylvania, in a universe not too
          different from our own, W. T.
          Warren has been hired to test out
          leather football helmets, which
          he's been doing for the past six
          months, when something peculiar
          happens.

Warren charges and leaps at the wall headfirst, but instead
of making contact, he glides right through the wall and
finds himself on the other side. He crashes into the
leatherworking equipment inside.

                    WARREN
          How in the hell...

                    NARRATOR (V.O.)
          Though the probability of such an
          occurrence is ostensibly zero, in a
          sea of infinite universes, it was
          bound to happen eventually. Most
          humans live their entire lives and
          only collide with a wall on average
          327 times. So, with 6,212
          collisions to his name, W.T. Warren
          was in the top 99th percentile for
          total number of wall impacts in a
          lifetime in the subcategory of all
          humans who have existed in the
          history of humankind.

Warren's head hits the wall over and over. SLAM! SLAM! SLAM!
Splinters and paint chips sprinkle to the floor as Warren
falls on his ass.

He gives a thumbs-up to the men watching from the sidelines.
They nod in approval and jot down some notes.

                    NARRATOR (V.O.)
          It only makes sense that he would
          have a higher likelihood than most
          people to experience quantum
          tunneling on a macroscale, causing
          his entire body to move through a
          seemingly solid wall. He rolled the
          dice more than the rest of us; if
          it was going to happen to any of
          us, then it was bound to happen to
          him.

CONTINUED: 3.

We see Warren's body move through the wall in slow motion. Every finger, every little hair.

Warren is on the other side of the barn wall again--his face is ghostly white.

> NARRATOR (V.O.)
> Of course he didn't have the context to understand what had happened to him. And without that context, we cannot blame him for his next course of action: to allow a man to plunge a five-inch blade straight into his gut at the local Penny Bazaar.

INT. PENNY BAZAAR - DAY

A group of seedy-looking men wielding knives surround Warren. Warren seems unfazed, until a blade sinks into his gut. The look on his face is less a look of pain and more of disappointment.

> NARRATOR (V.O.)
> Perhaps it was the 0.15% blood-alcohol level from the whiskey he used to cope with his inexplicable experience, or maybe it was the months of self-inflicted brain damage he incurred for what he thought was a respectable $1.25 a day. But most likely it was the little story his broken mind had begun formulating the moment he became aware of the fact that he had just jumped through a wall. You see, Warren had three great loves in his life: God, American football, and Fanny Lee Parker, the redhead who worked at the Marks & Spencer Penny Bazaar down the street.

MONTAGE: Vintage textbook illustrations of blood vessels, time-lapse X-rays of a brain experiencing months of trauma, images of biblical Renaissance paintings, stock footage of turn-of-the-century football games, and Fanny Lee Parker, the redhead who worked at the Marks & Spencer Penny Bazaar.

> NARRATOR (V.O.)
> He took up the position as a football helmet tester to fund his
> (MORE)

(CONTINUED)

CONTINUED: 4.

> NARRATOR (V.O.) (cont'd)
> expensive habit of watching Fanny
> as he pretended to shop.

Warren buys a scarf from Fanny just so his hand can touch hers as they exchange a penny.

INT. BARN - DAY

> NARRATOR (V.O.)
> So naturally, when he passed
> through the wall, his mind
> immediately went to Lazarus, and
> Peter, and stories from his
> afternoons at Sunday school.

MONTAGE: Illustrated images from the Bible of Peter walking on water and Lazarus rising from the dead.

> NARRATOR (V.O.)
> It was clear as a bell: God had
> chosen him as a vessel to carry out
> miracles to bear witness to His
> Mighty Power amongst the depraved
> and disheartened of his generation.
> And so when three armed men came in
> to rob the Penny Bazaar while Fanny
> and Warren happened to be there, he
> saw an opportunity.

INT. PENNY BAZAAR - DAY

Three men kick open the door and brandish knives.

> NARRATOR (V.O.)
> Warren knew with all his heart that
> God would allow the blade to pass
> right through his body.

A man plunges the knife into his gut.

> NARRATOR (V.O.)
> And to be fair, if it weren't for
> his L4 lumbar vertebrae, it
> probably would have.

Warren falls into Fanny's arms with blood flowing from the wound.

(CONTINUED)

CONTINUED: 5.

                    NARRATOR (V.O.)
          And as he slowly bled out, he
          thanked the Lord for this one final
          gift: a chance to be held in the
          arms of the woman he believed to be
          in love with. And while this might
          sound like a poetic, bittersweet
          ending to this story, it would only
          be fair for me to point out that,
          out of the infinite paths the
          universe took after that knife was
          plunged, there was a small subset
          of universes in which the knife did
          pass clean through, and Warren left
          that Penny Bazaar unscathed to
          continue on with the rest of his
          life.

EXT. WEDDING CHAPEL - DAY

Warren kicks open the chapel doors, Fanny in his arms.
There's the rice, and there's the tin cans tied to a car.

                    NARRATOR (V.O.)
          And if you think all of this feels
          like a fundamental misunderstanding
          of how the world works, then I'm
          afraid your opinion of infinity, my
          friend, may be too small for this
          story.
                (beat)
          To continue:

INT. SCHOOL HALLWAY - DAY

The camera pushes briskly past a long trophy display full of
plaques, trophies, and old photographs of teams from better
years. It stops on a dime in front of a giant state champion
plaque.

                    NARRATOR (V.O.)
          In the hallway of Prairie Heights
          High School in Prairie Heights,
          Indiana, is a solid mahogany plaque
          from 1957, when the Panthers, once
          the town's pride and joy, took home
          a win at the state championships.
          Beside it, in a framed clipping
          from the Heights Herald, Coach Hank
          Hendricks is quoted as saying,
          "There are two ways to catch a
                        (MORE)

                                                            (CONTINUED)

CONTINUED: 6.

> NARRATOR (V.O.) (cont'd)
> football: the right way, and the wrong way. My boys know the right way."

CUT TO: Newspaper clipping with Coach Hendricks's quote.

CUT TO: Close-ups of hands in two positions. One is the right way (open palms facing out in the shape of a triangle), and one is the wrong way (two hands twitching in the air in panic).

CUT TO: Photo of Coach Hendricks and his team. The camera comes to rest upon the face of one freckle-faced boy among his teammates: JIMMY FULLBERN (14).

EXT. FOOTBALL FIELD - DAY

Jimmy Fullbern runs as fast as he can to open himself up for a deep pass. The ball is thrown, and Jimmy reaches his arms up to receive the ball.

> NARRATOR (V.O.)
> Unfortunately for Jimmy Fullbern and the rest of his team, not even Coach Hendricks' quippiest of aphorisms was strong enough to stand against the inevitability that, in one of the many possible futures...

Just as Jimmy throws his hands up, we see the screen split into two--almost like a simple kaleidoscope--one in which his hands are up in the right way, and the other in which he is doing it the wrong way.

CUT TO: A low-angle close-up of a railroad switch being switched abruptly.

Cut back to Jimmy in the universe in which he is holding his hands up the wrong way.

> NARRATOR (V.O.)
> ...Jimmy catches the ball the "wrong way" in overtime at the state championships, not only losing the game for the whole team, but also receiving a nasty throat injury that prevented him from being able to talk for two whole months.

The ball slips through his fingers, slams into his esophagus with a CRUNCH, and right as he's about to hit the ground, we CUT TO:

INT. SCHOOL HALLWAY - DAY

The trophy display is suddenly empty. We hear the CROWD'S DISAPPOINTED YELL. The endgame buzzer BLARES, just as the lights in the hallway go out.

INT. HOME BATHROOM - DAY

Jimmy is unwrapping medical bandages from around his neck.

> NARRATOR (V.O.)
> Though no one would ever be able to tell the difference, Jimmy's larynx healed in such a way that he was no longer able to produce certain vocal frequencies. In many cases, this would not have mattered. But in the case of Jimmy Fullbern, most of the frequencies that would be forever lost to him just so happened to be in the same range of frequencies that the human ear has evolved to be most attuned to.

INT. HALLWAY - DAY

Jimmy is on the phone.

> JIMMY
> Yes, hello, I was wondering if Susan was in--

> VOICE ON PHONE
> What? Excuse me?

> JIMMY
> I was wondering if--

> VOICE ON PHONE
> I'm so sorry. Could you speak up?

> JIMMY
> Hello? Hello? Can you hear me now--

(CONTINUED)

CONTINUED:

                    VOICE ON PHONE
          What?

                    NARRATOR (V.O.)
          This made most situations in which
          he had to speak a great
          inconvenience.

INT. SCHOOL DANCE - NIGHT

Jimmy is surrounded by dancing young couples. He taps the shoulder of a young girl. She turns around.

Jimmy yells over the music but is still completely drowned out. The girl attempts to understand but eventually gives up.

                    NARRATOR (V.O.)
          But in many others, downright
          devastating.

Jimmy watches as the girl walks away.

INT. WORKSHOP - DAY

Jimmy is older now, sanding down a table.

                    NARRATOR (V.O.)
          Seeing as no one cared to listen to
          what he had to say, Jimmy learned
          to get by without saying much. He
          became a carpenter because he was
          good with a saw and appreciated the
          fact that tables didn't talk.

Jimmy blows away sawdust with great care.

                    NARRATOR (V.O.)
          Though I know for a fact that, in a
          universe where tables do talk, he
          lived a far less lonely life.

Jimmy looks down at the table lovingly.

                    TABLE
          That was really good. But I might
          need a little more on the left.

Jimmy smiles and nods, and he continues sanding.

(CONTINUED)

CONTINUED:

> NARRATOR (V.O.)
> And he wouldn't have felt the need to hang himself from the workshop's rafters.

CUT TO: Jimmy's legs dangling above the table.

> NARRATOR (V.O.)
> I know you may be saying to yourself, "If only, if only." But I wouldn't waste another breath on that line of thought, because this is also where he would be had he caught that ball, those many years ago.

CUT TO:

INT. SUPER CHIC '70S LOUNGE - DAY

A hundred dead cult members lie across shag carpet and leather couches. They all wear matching white button-down shirts and khaki pants.

In the middle of the pile is Jimmy Fullbern, also dead.

> NARRATOR (V.O.)
> You see, it just so happens that, had Jimmy caught the ball the "right way" back in 1957, those perfectly normal vocal cords paired with his dangerously persuasive brain would have made him irresistible to anyone he happened to speak with.

CUT TO: Old 16mm footage of Jimmy preaching at a pulpit and emphatically waving his arms. His followers close their eyes and nod up to the skies. Some of them even have their hands up, palms out, in the shape of a triangle.

> NARRATOR (V.O.)
> Coincidentally, the "right way" to catch a football just so happens to bear a striking resemblance to the Illuminati hand sign of the Great Pyramid that Jimmy would eventually adopt as the official hand sign for his own cult. Though Jimmy never made the connection himself, this fact would have tickled his fancy.

(CONTINUED)

CONTINUED: 10.

MONTAGE: Different cult members using the hand sign in old archival photos.

BACK TO: The pile of dead bodies in the living room.

> NARRATOR (V.O.)
> And so, after all of this fussing about "right" and "wrong," none of it really seemed to matter in the case of poor Jimmy Fullbern.

EXT. BLACK VOID

> NARRATOR (V.O.)
> And then, there is the case of you.

In the black abyss of space, we see swirling galaxies.

> NARRATOR (V.O.)
> In the tiny subset of universes in which life has had the pleasure of coming into existence...

The camera zooms quickly to Earth.

> NARRATOR (V.O.)
> There is a tinier subset in which this book also exists, and only a .00003% chance of it then making its way into your hands...

As the camera zooms in, we find you holding this book.

> NARRATOR (V.O.)
> Making it exceedingly unlikely that you'd read these deleted pages from an early draft of a screenplay as though glimpsing an alternate universe and, upon reading them, shrug a bit, confused by it all.

You glance around unimpressed.

> NARRATOR (V.O.)
> Wow. You're witnessing a miracle here. This rare moment alone makes every dollar you spent on this book worth it. So you should consider everything that happens from here on to be a bonus ,and feel grateful.

SLOW MOTION: You turn the page...

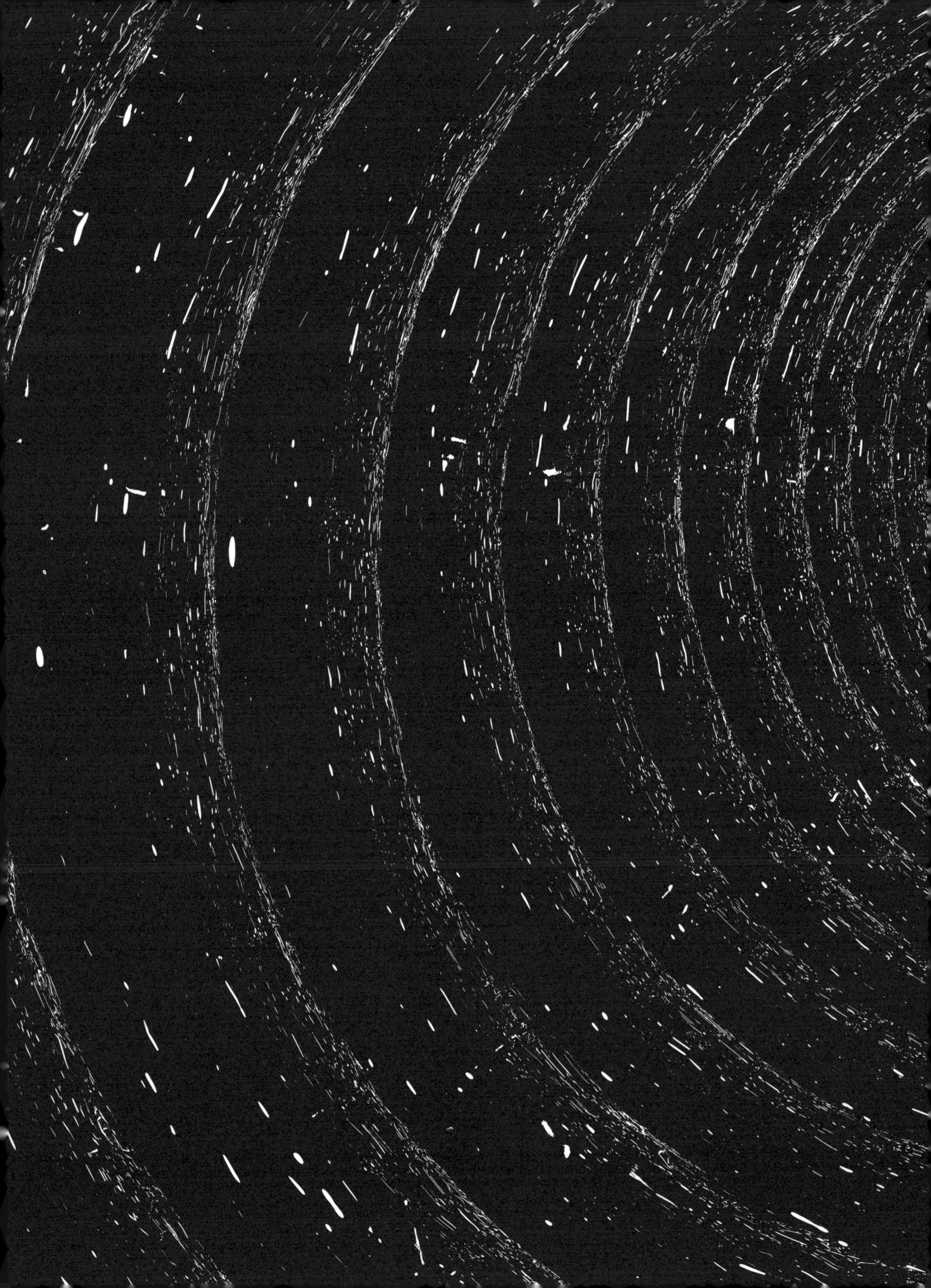

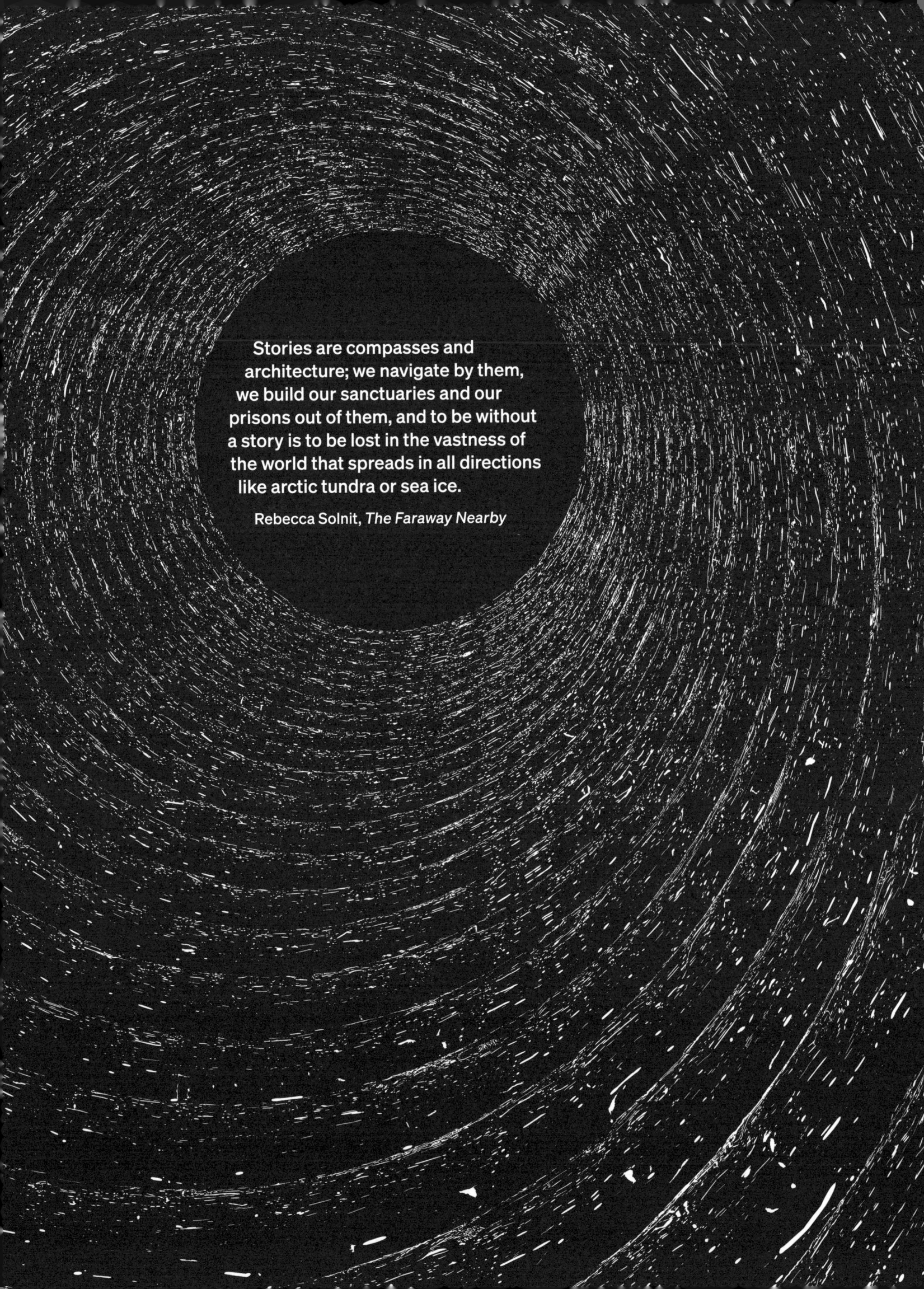

> Stories are compasses and architecture; we navigate by them, we build our sanctuaries and our prisons out of them, and to be without a story is to be lost in the vastness of the world that spreads in all directions like arctic tundra or sea ice.
>
> Rebecca Solnit, *The Faraway Nearby*

> I toss the stone of my story into a vast crevice; measure the emptiness by its small sound. —Carmen Maria Machado, *In the Dream House*

Chapter 1

Parallel

Quantum theory tells us that every time we observe the world—observe where an electron is inside a box, say—we see only one of multiple realities that had been possible the moment before. There's no way to predict what we'll observe in advance, no preset evolution of the universe as time goes on, no inevitable future. Quantum law only lets us speak of the probability of different outcomes. Then, at the critical moment, one outcome manifests at random. The electron shows up on the left side of the box, say.

Physicists have grappled with the meaning of quantum mechanics for a century. The standard view, if there is one, traces back to quantum pioneers Niels Bohr and Werner Heisenberg. Their "Copenhagen interpretation" accepts that the universe is random and indeterminate—that the instant an electron's location is measured, it's positioned at a point in the box, as if by the roll of a die.

Today, however, most quantum physicists rate the "many-worlds interpretation" —the brainchild of midcentury American physicist Hugh Everett III—as equally consistent with experiments. Everett argued that all possible measurement outcomes materialize. When you measure an electron's location, for instance, the universe branches into multiple realities: one in which the electron turns up on the left, one where it's on the right, one where it's dead center, and so on. Particles interact, thereby "observing" each other, constantly, so the universe splits into more and more versions of itself at an incomprehensible rate.

These proliferating realities run in parallel, unable to affect one another. The reality in which you saw the electron on the left and the reality in which it materialized on the right continue to split, each engendering countless "daughter" universes that themselves spawn whole branching trees of parallel universes.

Thinking about all possible realities playing out at once is cognitively straining, even for physicists. Some deem the many-worlds idea comically unparsimonious, while others consider it the most straightforward and complete interpretation of quantum mechanics. Any quantum physicist who truly believes that every quantum possibility plays out should have no qualms about playing a particle version of Russian roulette, where, for instance, detecting the electron on the left side of the box pulls the trigger of a pistol, and detecting it on the right does not. According to the many-worlds hypothesis, you're guaranteed to survive in some reality either way. So far, no physicists have taken the gamble, but the many parallel worlds might exist all the same.

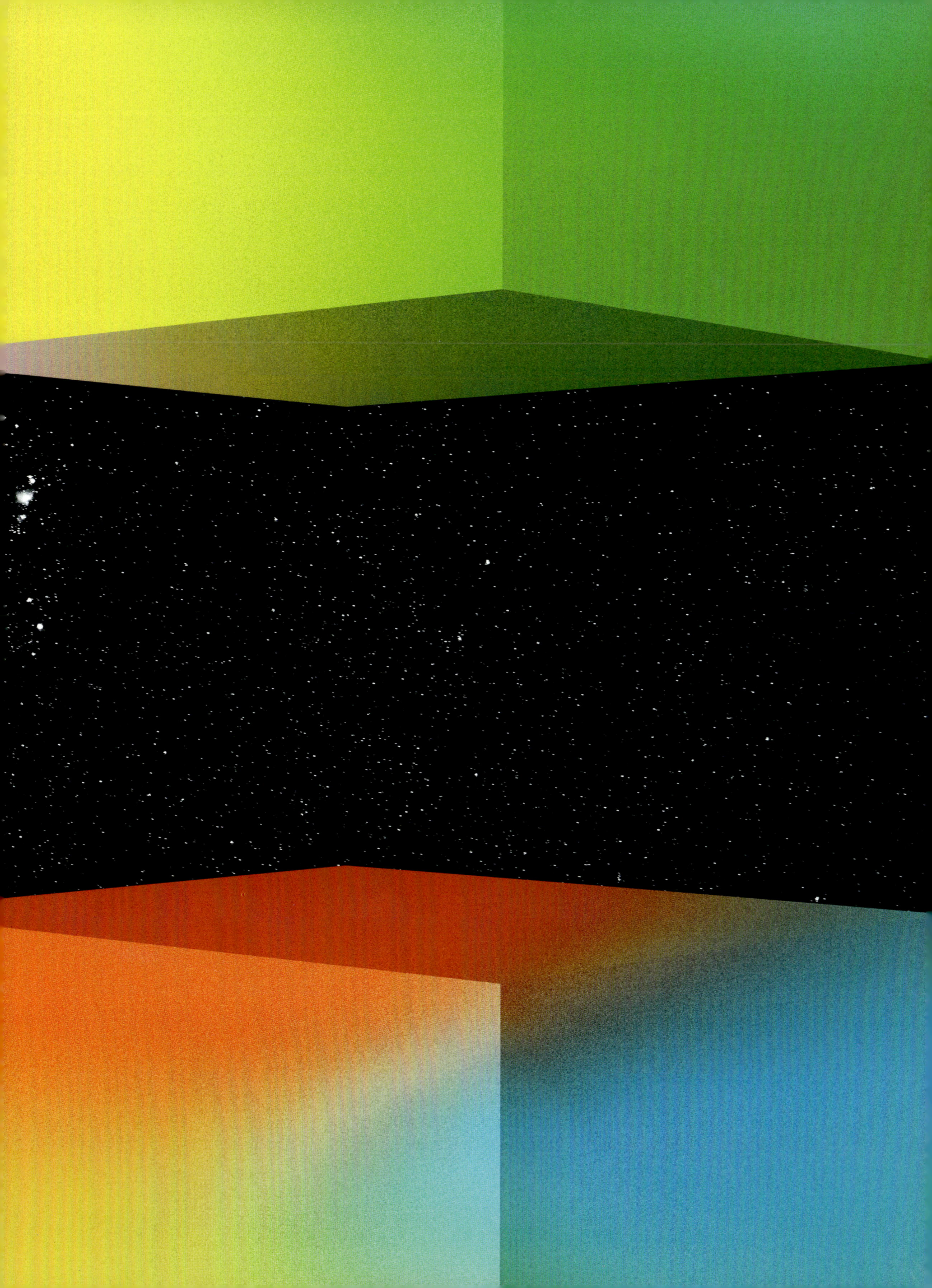

It is in the highest degree unlikely that this Earth and sky is the only one to have been created [...]

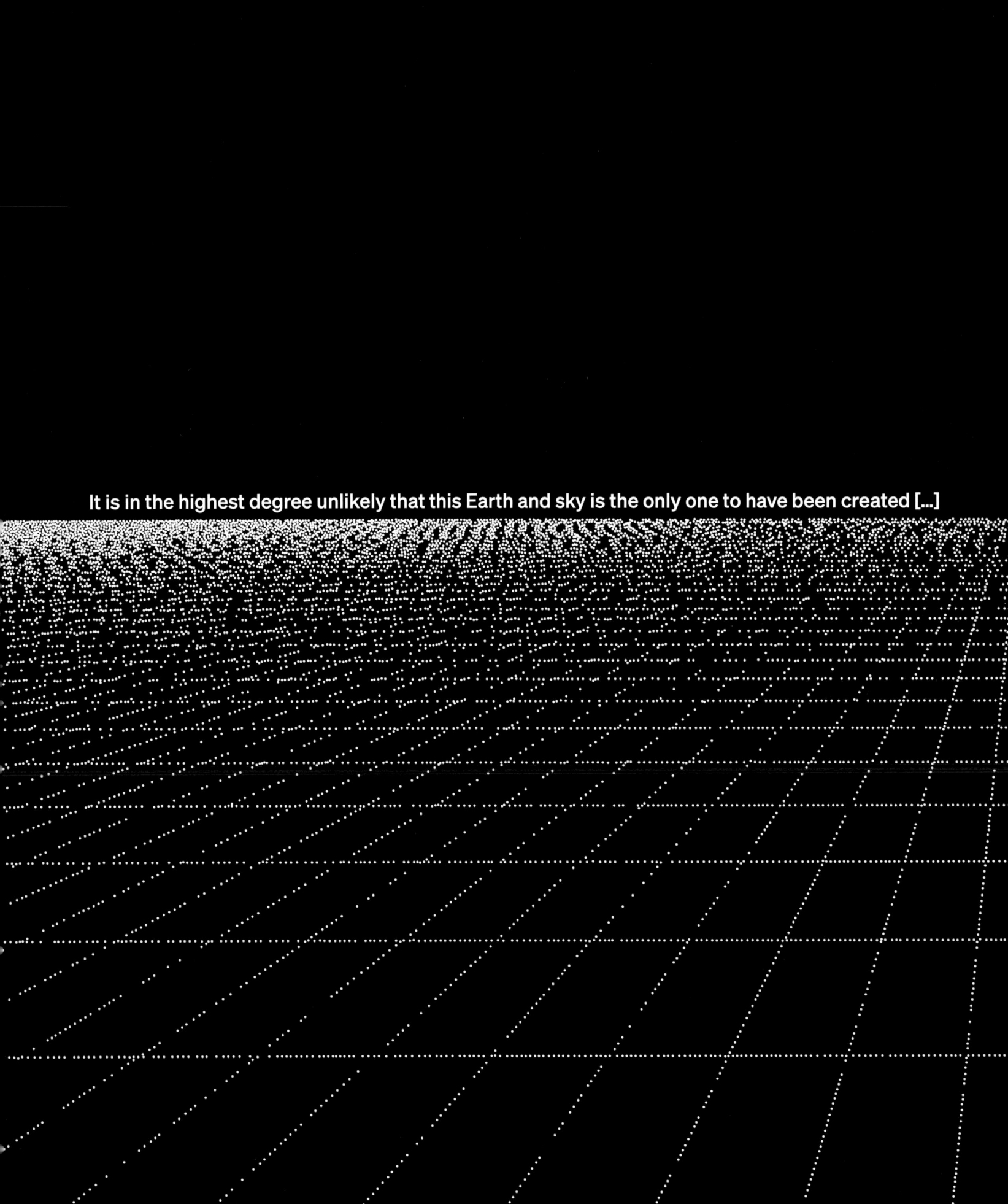

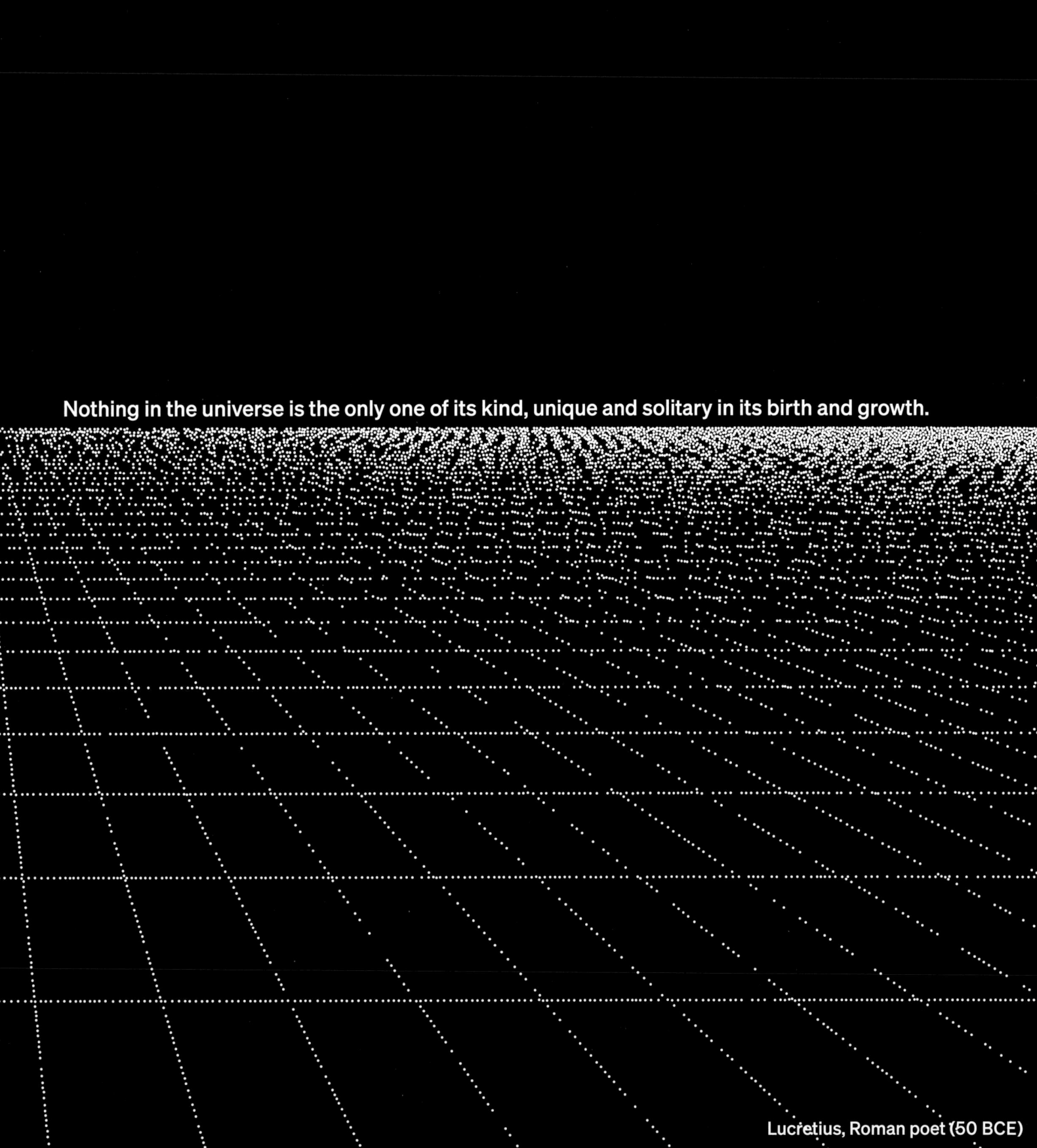

Nothing in the universe is the only one of its kind, unique and solitary in its birth and growth.

Lucretius, Roman poet (50 BCE)

IT
WAS
ONLY
BROKEN
IN
TWO
PLACES

CHARLIE + RYDER

Shock

by Esmé Weijun Wang

It was only broken in two places.

Ryder kept repeating the fact, like a poisonous mantra, as she and Charlie hurtled toward the hospital. Charlie thought he would punch his wife if she said it again, and yet Ryder kept saying it. After every instance, bringing up the notion of a still more tragic fate for their son, Charlie flinched—as an indoor kid, he had never broken a bone, but Ryder was a former gymnast. Being an athlete made life fundamentally different. Hadn't Kerri Strug proven that in the 1996 Summer Olympics? But their twenty-year-old son wasn't an athlete; moreover, Jonathan had jumped out of a window. *That* was the fundamental difference.

"He could've *died*," Charlie finally snapped. They were close to the hospital now; he recognized the neighborhood as the one he visited to pick up Jon's medications.

Ryder replied, "He could've, but he didn't. I'm just saying that it could have been worse."

"Very optimistic of you."

"A broken leg will heal. He's alive, and now he falls under the 5150."

This was true. They hadn't been able to force Jon into the psychiatric ward, but now that he had actively attempted suicide, he could finally be committed. Charlie's fist loosened a fraction. He scratched absently at his face.

At the hospital's front desk, the receptionist took their names and made calls until she reached someone who was able to report that, because Jon was still in the ER and waiting to be transferred to the proper ward, they could not see him until he was transferred.

"Is he in restraints? I know that's why we can't see him," said Charlie.

"It's *fine*. We're not going to throw a shit fit about restraints."

"What he means," Ryder quickly said, "is that we know the drill."

The nurse stared at them. "I'm sorry," she said, "but that's the policy."

"Well, when is he going to be transferred? How are we supposed to know when he's been transferred?" Charlie asked.

"They'll call you," said the nurse.

Charlie and Ryder returned to the waiting area. The land of plastic orange seats was far from full—a weekday night, Charlie supposed, would lead to fewer medical emergencies—and some of those who were waiting seemed to be in agony. An elderly man moaned to himself as he curled forward, like a tender fern, over his knees. A woman held a screaming baby to her chest. Everyone looked raw and in need of care.

"I don't think they have our phone numbers," Charlie said.

Ryder said, "They'll just call our names, babe."

"Fuck," Charlie said, and the tears came in spite of himself. Ryder wrapped her arms around him, her hand rubbing his back in a manner not dissimilar to how she had rubbed Jon's back when he was a baby, when she bounced him up and down in her arms. Engulfed, Charlie could smell the deodorant she used—something apricot-scented and sweet. Before they had Jon, he had never cried in public, and now, with Jon in his early twenties, Charlie had cried in public and private far more than he felt was appropriate.

Ryder, on the other hand, seemed to have hardened in response to distress. If one of them was upset, the other had to compensate by being stoic. It was an unspoken rule in their marriage, as sacred as any vow made in the Catholic church Ryder had insisted on for their wedding, and for years, neither of them had broken the covenant. But they were trapped now—calcified: Ryder was strong, while Charlie couldn't help but crumble. Now she pulled away from him. He shrugged, embarrassed.

•

When they finally saw Jon hours later, he seemed less like their son and more like a Jon who had been eviscerated and stuffed with straw. They met in a corner of the ward during visiting hours—Charlie and Ryder had been made to wait, so as to not disturb the other patients' routines—and Jon, their beloved, clever son, had been dulled at the edges. Charlie wanted badly to hug him, but he and Ryder had been told at the outset that touching wasn't allowed, so instead, Charlie clapped his hands together, soundlessly and repetitively, to give himself something to do.

"I'm sorry you had to come out here," he said.

"No, baby," Ryder said. Her hand rose to his shoulder, hovered, and then swiftly withdrew. "You don't have to apologize."

Charlie couldn't stop staring at Jon's leg, which was partially why it had taken so long to get him into the ward. Because of the nature of the breaks—multiple and compound—he had required surgery to implant an internal fixation device, which they would remove after six to eight weeks. It looked monstrous, with its metal scaffolding and screws. "Does it hurt?" Charlie asked, gesturing.

"My leg..." Jon slowly looked down.

He was on Zyprexa, which always made him robotic; Ryder would fight for him to be transitioned to another drug. "It's broken."

"It seems pretty bad," said Ryder. Her hand came up again, and this time, she quickly squeezed Jon on the shoulder before returning her hand to her lap.

I should've thought of that, Charlie thought. Just be fast.

"You'll be here for a bit," she continued. "Until some things get straightened out. They'll help you."

Silence. Both Jon and Charlie were still gazing at the broken leg.

"I can't be helped."

"What?" Charlie's head jerked up, as if controlled by string.

"It's true."

"It's *not* true," Ryder said, and her hand came up again until a loud voice asserted, "No touching," and her hand dropped without any of them turning to look at the nurse who had said it.

"Mom's right. You'll get better," Charlie said.

"No," Jon said, "I won't."

•

Ryder stood at the kitchen counter, stirring her tea with a spoon. She'd used three bags of PG Tips tea to make it extra strong. She dripped honey into the cup. She had showered to get the scent of sex off. Rubbing deodorant into her armpits and blow-drying her short hair, she was still exhausted from the afternoon's exertions, and Charlie would be home in a few hours, which required a certain amount of alertness. He would be tired and grumpy. She would cheer him up. That was how their marriage operated. It was stupid, really, to want anything else but a casual fuck now and again; like Jon, Ryder required an occasional shock in order to return her to life. Someone had told her—one of the other mothers in the PTA, someone whose name she couldn't remember—that having Jon would be the best thing she ever did. She pressed her mouth hard to the mug's rim. The liquid scorched her lips and she kept them there.

Broken

by Etgar Keret

Translated by Jessica Cohen

It was only broken in two places, and the doctor said the bone would be fully healed within a few weeks. With a little physical therapy and some exercises, three months from now—four, tops—Ryder would barely remember the accident. "How many times have I told you?" Charlie grumbled while Ryder was getting her cast put on, "How many, huh? Don't ride your bike when the sidewalk is frozen!" "Sir," said the doctor in a stern voice, "would you please stop? This is not helpful." When Charlie asked, "What isn't helpful?" the doctor lost her patience and asked him to wait outside. "Did you see the way that bitch threw me out?" he said to Ryder when they were sitting in the ER waiting for their discharge forms, "Like I was some kid talking back in class?" "It's just that she could see how much pain I was in and how hard it was, and she didn't want th..." Ryder fell silent midsentence. Charlie was also quiet for a moment. "I'm sorry," he finally said, "You're right, I shouldn't have attacked you like that. It's just that, when I see you hurting, I think, hell, I'd do anything to make it stop. And when I can't, it just makes me want to smash something. You get me, don't you, babe?" Ryder nodded. "It's like...," Charlie continued, "it's like there's some parallel universe, where you listened to me and you didn't skid with that goddamn bike. And in that world, instead of waiting around to get discharged from the ER, we're having sex on the kitchen table, or eating Cherry Garcia and watching some lousy series on Netflix. And I'm jealous. I'm jealous of that other Ryder and Charlie who live in that parallel universe, and I wish I could switch places with them and..." He hesitated as if he wasn't sure how to go on and then abruptly stood up: "I think that tight-ass receptionist is back at his desk. Wait here, okay, babe? I'll just be a minute." While Charlie took the forms over to the young guy with the eyebrow piercing, Ryder tried to imagine Charlie's parallel universe. She pictured the two of them eating ice cream in front of the TV, or fucking in the kitchen. Or maybe all at once: watching Netflix while fucking and eating ice cream. She could see Charlie arguing about something with the guy. She shut her eyes, and for a moment she could practically taste that ice cream on her tongue.

•

It was only broken in two places, and Charlie immediately apologized and said it could be fixed. "Two drops of glue for three seconds and it'll be like new," he said in a trembling voice, and quickly added, "I'm sorry, I just lost it there for a second..." Five minutes earlier, Ryder had come home from yoga. Her class was on the other side of town, but she'd decided to walk instead of taking the bus. It took her almost 50 minutes, and halfway there, she realized she'd forgotten her phone. By the time she got home, Charlie was flipping out. "Your class ended an hour ago!" he yelled, "I've been calling you for a whole fucking hour and you didn't answer! Do you know what was going through my mind? I thought you'd been mugged, or murdered, or ra—Forget it, I don't even want to say it out loud." Charlie flicked his wrist sharply, like he was slapping someone invisible with the back of his hand, and knocked the blue vase off the table. It was a housewarming gift from Ryder's grandmother. They both froze, and then Charlie bent over and picked up the pieces. "I'm sorry, babe," he kept mumbling while he dug through his toolbox for a tube of super glue. When he found it, he grinned and said, "I'll have this fixed in no time. It'll look brand new. Better than new." He put the ceramic pieces on a sheet of newspaper and breathed in dramatically, like an Olympic swimmer about to dive into the pool. They both knew that with his two left hands, even if he managed to glue the vase back together, it would always be cracked and look it. They also knew that this exhausting ritual was unavoidable, because Charlie was intent on showing Ryder and himself how remorseful he was. While he applied a thin strip of glue with a Q-tip, he said to Ryder, "I screwed up. I got stressed 'cause I thought something happened to you and I screwed up. You know I bought us ice cream and everything for tonight? That cherry flavor you love. And I thought we'd watch that show about the Korean English professor that Tom said was really funny. And now…" He looked at the pieces of pottery on the newspaper and tried to figure out how they fit together, but after a few moments he gave up. "That's the last thing I got from Grandma before she died," Ryder murmured, and Charlie got up to hug her. "I know. I'm sorry. I know… You know what I can't stop thinking about? How there must be some parallel universe where, instead of flipping out and breaking things, you and I are together on a desert island, with no blowups, no stress, no freezing cold weather, no broken vases. Just you, me, a gorgeous beach, and some tropical trees… Can you imagine something like that? I can literally feel the sun warming my face."

•

It was only broken in two places. Charlie shook the coconut over his mouth, but nothing came out. He picked up a rock and bashed the coconut again, as hard as he could. The rock struck his thumb and he screamed in pain. Ryder said he should put something cold on it to stop the swelling, but there was nothing cold anywhere nearby. Only golden sand, lush coconut trees, and a spectacular, endless ocean. "Maybe dip your hand in the water?" she suggested, but Charlie grabbed his throbbing left thumb with his right hand and yelled, "Why did we even get on that sinking boat? Why did we have to waste all that money on a lousy sailing tour? Just think, we could have been in Middletown right now, in the park across the street from home, building a snowman, having a snowball fight, goofing around together like kids. Hey, babe? Could it be any better than that?" Ryder didn't answer. She just looked up at the blue sky, and instead of thinking about Middletown and Charlie and the snowman, she savored the warm sun caressing her face and the monotonous crashing of the waves.

Daybreak

by Emily Segal

It was only broken in two places. The upper left corner, then closer to the tab that locked the car door, where the window was tinted darkest. Broken was perhaps an overstatement; the girl in the back seat had managed to merely fracture the glass with her fantastical, brutal flailing. The glass itself was still intact, sealing off the car's inhabitants from the rest of the world, as they slid through the streets near the Lincoln Tunnel. It was not yet dawn. Soon, Charlie, who was driving, would see a smear of orange on the horizon. Ryder, in the passenger seat, was dabbing at the thigh of his white canvas work pants, where his milky bodega coffee had spilled in the fracas. "YOU FUCKING FAGGOTS!!!" the girl screamed in the back seat. "PULL OVER. RIGHT FUCKING NOW."

Ryder glanced over at Charlie, the cooler cucumber. Archer Fast was the most high-profile child they'd been paid to "onboard," as the materials so coolly put it, and even though her parents were entirely in on the scheme, it still made him uncomfortable. Wasn't she, like, American royalty? Ryder said this without saying it, his gaze flickering over the stubble of his partner's chin. Tiles appeared on either side of them at the mouth of the tunnel, which was jammed.

"Fuck," Charlie said, "this always fucking happens." Then, to Ryder, "Open that thing up. She's wiling."

As if to illustrate, Archer beat her forehead against the back of Ryder's seat, knocking his blue-and-white Greek coffee cup, milky micro-ocean swelling, the whites of Ryder's eyes visible. Exasperated, Charlie reached over and opened the glove compartment himself, withdrawing a black neoprene, lozenge-shaped case with one hand while blaring the horn with the other. Ryder thought of his parents, who'd never failed to mention the film scene in which the tiles started popping off the walls of the Tunnel, flooding it. Charlie, as usual, seemed to think of nothing but forward motion.

"Open this." Ryder unzipped the case, which held a massive needle filled with fluorescent orange liquid.

"Looks like Tang today. A Tang lava lamp." Ryder often attempted levity at strange moments, a habit that irked Charlie. The car edged into the murkiness of the tunnel, where traffic was starting to move. Ryder undid his seatbelt, twisted and reached for Archer, syringe between his fingers. Without thinking, he went to cover her eyes; she reeled back and bit him—hard—but not so hard that he couldn't get the needle into her neck.

Parallel

Before the pain rose, there was only heat and a sense of connection, and Ryder circuited to this pale face, with its summer freckles fading, dirty blond hair under a huge hoodie from what must have been her school, green with a white tree embroidered on the chest. Maybe it was the familiarity of Archer's face that made this weirder, harder to bear. He'd seen it lined up with the rest of her family in the checkout aisle at D'Agostino's, shining from the cover of *People* magazine. None of the other kids they picked up in the night for "onboarding" had rung a bell. They may have been rich, but they were anonymous.

"She bite you, bro?" The traffic was moving more quickly now. "Get that wrapped up." Charlie glanced at Ryder and then back at the road, his voice tinged with impatience. At this pace, they would almost definitely be late to Bishop Ranch for Troubled Teens. The coffee stain on Ryder's pants was now bordered by red, turning brown, and the heat where her teeth had hit was turning fervid. He took off his hoodie and wrapped the sleeve around his hand, tight, double-wrapped it.

"She's out."

"Famous little cunt."

"What'd she do, anyway?"

"Blow. Always blow for those little private school girls." Eavesdropping in the office at the ranch, Ryder had gathered it also had something to do with a particular photograph, now popular on the internet, of Archer spitting gleefully into the mouth of another girl. Sunrise flooded the car as they exited the tunnel. A supermodel glinted from the David Yurman billboard past the tolls. Silver helices and grayscale ocean water. If Archer had been conscious, she may have noted it as a boundary marker, the imprint of the world she was leaving behind: the Upper East Side.

It was almost funny how humorless Charlie was about it, stripping down in the locker room after a long drive from the city, "escorting" a "new recruit" into "onboarding," prepping for "campfire time." He always folded his clothes carefully, Carhartts sectioned three times, laid on the top shelf of his locker. He urged Ryder to be more masculine, more brutal in the way he subdued the recruits. Backstage, during orientation, Ryder always wondered if the kids would remember him in the future. Would they know, however obscurely, that he'd been the one who zip-tied their wrists while the world slept, mommy whimpering in the hall that it was for the best?

For the most part, Charlie seemed immune to these kinds of questions. It was only on their early morning drives to camp with their catatonic charges that Charlie ever showed hints of a more fanciful, even romantic side. Watching the sun rise over the Jersey Turnpike, he loved to take a particular turn and piss by a meadow. They took this route today as well. "Magical, magical meadow," Charlie murmured. Birds squawked, cirrus clouds streaked the sky, engines roared past. Archer immobilized, zip-tied, raw-wristed, drooling in the back seat.

Charlie whistled as he emptied his bladder. Peering out, Ryder depressed the latch of the glove compartment, slid his hand in, eyes still on his partner. Perhaps today they would take a more circuitous route.

Gravel crunched as Charlie whistled his way back to the vehicle. *Tang-a-lang*. The beep of the sensors as the door opened, the slam-hiss of the door swinging shut, then a pop, as Charlie's skin broke, with only the briefest of squeals.

BILLY CHEW

All night, no matter how many times Lily rethought and restrung the twine connecting the family photos and old news clippings pushpinned to her bedroom wall, there were always breaks in the web, abrupt dead ends, holes in the story her grandmother had always told her.

Which meant either that Gran had been lying to Lily for thirteen years or that even the family matriarch had no idea that Uncle Finn was still alive in Tucson.

When she finally found the gravestone of her late grandfather, it was covered in offerings of jam preserves and sheet music, though Lily had found no record of him liking either.

Demoralized, she began to walk away, knowing she would have to start over again in the morning, but as she turned to have one last look from the doorway, she felt her knees go weak: the twenty-seven breaks formed a distinct shape she had not seen in years.

It all backfired, and the burial of Motorboat wasn't even the latest in a catalogue of 27 heartbreaks, the freshest of which was a sanitation worker named Yussef who broke it off with her yesterday to "dedicate more time to his Art."

JULIA POTT

She hadn't meant to date 26 people at once, but her dachshund had just died and she needed the distraction.

However, using 26 people as transitional objects doesn't really work when they all have their own personalities and agendas, and when none of them wanted to come to Plum's memorial, which was held near Plum's favourite puddle down the street from her duplex.

Unfortunately, this time, it was broken in 27 places.

Harry took the curious piece of her heart that wanted nothing more in life than to read books and share fun facts with someone special; Shane took the ambitious part of her that believed she could change the world; Oliver, Ethan, and Liz all took the parts of her that would make plans with friends but then get high, forget, and instead stay up and get late night Pizza Hut; Lee, her superstitions; Frankie, her kinks; Joshua, her bird-watching hobby (along with her favorite pair of binoculars); Jack, Samuel, Meghan, Reinhardt, Suni, Annette, Ajay, Hakeem, Josiah, Joe, Joe #2, Zion, Chin-Hae, Bartley, Butters, Daniel (or was it David?), Dwayne, and Christoff, the part of her that wanted to fuck—leaving just one piece for her to call her own.

So when I realized I had nothing to lose, I confidently leaned into the microphone and began to tell everyone what I truly thought of the organization.

DANIEL KWAN

It was too late—the audience had seen what they could not unsee, and I knew by the end of the presentation, I would be out of a job.

I'd only started working here because of nepotism anyway, but I'd grown accustomed to marine biology. And spending time in the ocean brought me closer to my sister, who died in a fishing accident when we were eight.

Seeing the veins in Pete's neck swell as he attempted to wave me off the stage, I realized I had only one option: keep the presentation going for as long as my body would physically allow.

Although overwhelmed by what lay ahead, Lily dug Uncle Finn's decade-old goodbye letter out from the bottom of her dusty old toy box and slipped it into the front pocket of her packed suitcase.

She had found him in the phonebook but was reluctant to call. She'd never forgiven him for beheading her rabbit when they were kids, but if calling him was what it took to get answers then... No, nope, sorry, I can't do it. He's such an asshole.

As she watched the dry Arizona dust beat against her tiny room's single barred window, she sank to the floor and finally accepted that her bastard uncle was right: he had abducted her over a month ago, and they were never going to look for her here if they thought he was already dead.

A voice in the back of her mind whispered, "Bring the apricot preserves to choir practice tomorrow morning and use them to reward your friends for singing the incantations on the score left here for you by my acolytes, my dear granddaughter."

The sheet music had played Judy Garland's devastating hit "The Man That Got Away," and as she ate the unsweetened preserves, she realized Grandma was trying to tell her something: that the night had been bitter with Grandpa and sweeter with someone else.

Her heart broke as she looked up at the web on her wall and realized no one in her family actually knew the real granpop, at least no one as much as the man who would spend every Sunday with him, picking peaches and practicing Vivaldi's "Concerto for Two Violins."

She could just barely make it out, but gradually, it scrambled into her awareness that she was staring at the New York City subway map, somehow color-coded perfectly.

The shape of the baseball field awakened memories of her grandmother's ghost, a former baseball announcer, who used to roam the hallways of her house, demanding that everyone "play ball." Lily spoke into the ether, "Okay, Grandma. I'm ready to play."

Whether it was an intentional message from her grandmother or just the blood clot forming in her brain causing her to find meaning where there was none, it didn't matter to Lily, because she knew in her heart what she needed to know.

His art was boring, hopelessly so, and she realized when he left her apartment that she'd probably told him that, albeit obliquely, one too many times.

She didn't believe in art, not after everyone had to move to Kepler 186-f and couldn't bring any art supplies, and everyone's space art was just deeply superficial musings about what had happened and what we'd left behind, and the art only served to take more time away from dating her.

Fortunately for Yussef's "Art," an artist's work tends to increase in value after they die, and she was left with no choice but to do to Yussef what she had to do to Motorboat and any of the others after she caught them running away.

I saw her there weeping alone, and when I approached, she wiped her nose on her sleeve and snorted, trying to politely smile hello through streams of mucus.

She moved into the puddle shortly after the memorial, though she'd auctioned her Wellington boots on eBay the week before in a desperate attempt to clear out the debris of her life, so her socks were wet and her nose was wetter, but she never felt closer to Plum than she did with wet paws in a puddle near the house she used to live in and Is that a squirrel? I'm just going to go check it out, just for a minute... I'm not going to hurt it.

I watched from my window as she awkwardly attempted to pour out Plum's ashes from an old shoebox while filming a video with her other hand, accidentally spreading remnants of burnt dog all over her new beautiful dress; we are all truly alone.

As she stumbled through the wreckage of her life, searching for whatever the hell that surviving piece of her was, she thought for a moment of the 9/11 first responders who spent weeks searching the rubble of the World Trade Center for people crying out, somehow still alive amid the chaos and carnage, and as she laughed to herself at the offensive comparison, she uncovered the remaining piece of herself buried beneath the debris: her bleak sense of humor.

The piece of her innermost toe that ached just before the hurricane hit her home.

The only part of her left was the part of her that compulsively breaks things into smaller parts, so she steadied her breath as she opened Tinder and prepared to break the one last piece of her into even more tiny pieces.

At least that's what I intended to do, but for whatever reason—perhaps it was the adrenaline—I committed to a lie: "Fine. The prototype is irreparably broken. However, my fellow pieces of shit, I'm pleased to announce that additional bomb prototypes have been planted beneath each of your seats here in this function room, and if anyone stands or attempts to follow me out of this room, those explosives will detonate, and you'll all get a taste of the karma you so badly deserve." And with that, I exited stage right—shouting something about being an angel of righteousness and retribution over my shoulder, as I threw open the fire exit.

"A cornflake mozzarella bar is not a thoughtful addition to the already oversaturated elevenses* market, as cornflakes wilt when marinated in cheese, and the reason I never shared my contempt for this flawed endeavour is that I am turned on by the stench of people's blind, unearned confidence in niche business ventures." (*In England, elevenses refers to a morning snack eaten at 11 a.m., usually consisting of tea and a biscuit. In New Zealand, elevenses are eaten around 10:30, and in Hungary, they eat Tíz-órai, which means "of the 10 o'clock." In the United States, it is simply known as a coffee break, and an aubergine is known as an eggplant.)

"This company is filled with nothing but liars, backstabbers, and ass-kissing sociopaths, but I have never felt more at home, and I will truly miss you all." I stomped on the prototype one last time just to make sure—even if everyone watching hated me for undoing years of work and millions of dollars in development, at least they would never forget me.

It seemed now that I'd soon be joining her in the afterlife, my imminent job loss suddenly irrelevant, as the 27 cracks in the Australian saltwater crocodiles' enclosure began to spiderweb and creak, the tank's water pressure quickly giving way to an explosion and tidal wave of hungry, angry, 20-foot, 2000-pound reptiles, all of them spilling, thrashing, and chomping into the Animal Encounter Paddock at the Perth Children's Educational Zoo & Reptile Rehabilitation Center.

There aren't a lot of days that I wish I wasn't an octopus, but this was certainly one of them, and when all but one of my hearts had stopped beating and my chromatic chatter was coherent only to me, I wondered who would possibly know me again—like my sister once had and like my colleagues had pretended to.

I couldn't help but think of her, as I looked from the screaming families sitting in the splash zone down to the orca whale that had foolishly trusted me to guide him through his trick routine but now laid contorted and gasping for air from the broken ribs that had punctured his lungs in the accident: "I'm not leaving you this time."

Unfortunately, however, because I already had to pee at the start of the presentation, I could only keep filibustering my termination for another thirty minutes, and when I tried to discreetly relieve myself into the Starbucks cup resting inside the podium in front of me, it made the whole situation considerably messier.

My right finger set alight first, flickering gently on the podium, and as I tried to put it out with my now entirely afire left hand, I realized how ironic it was that I had been giving a presentation on neur-asthenia and its applications in studying modern-day burnout while I was literally burning to the ground in front of Pete.

I picked up as many of the pieces of the "military-grade," "virtually unbreakable" lemonade pitcher as I could, squeezing them together in my hands before the audience, as if to showcase a product feature that would miraculously allow the pieces to come back together again. I squeezed tighter and tighter, blood began to trickle down my forearms, and the audience leaned in filled with antici-pation, as we all waited for a magic trick that wasn't coming.

A CAT   A INTERS

ECTION   PEOPLE

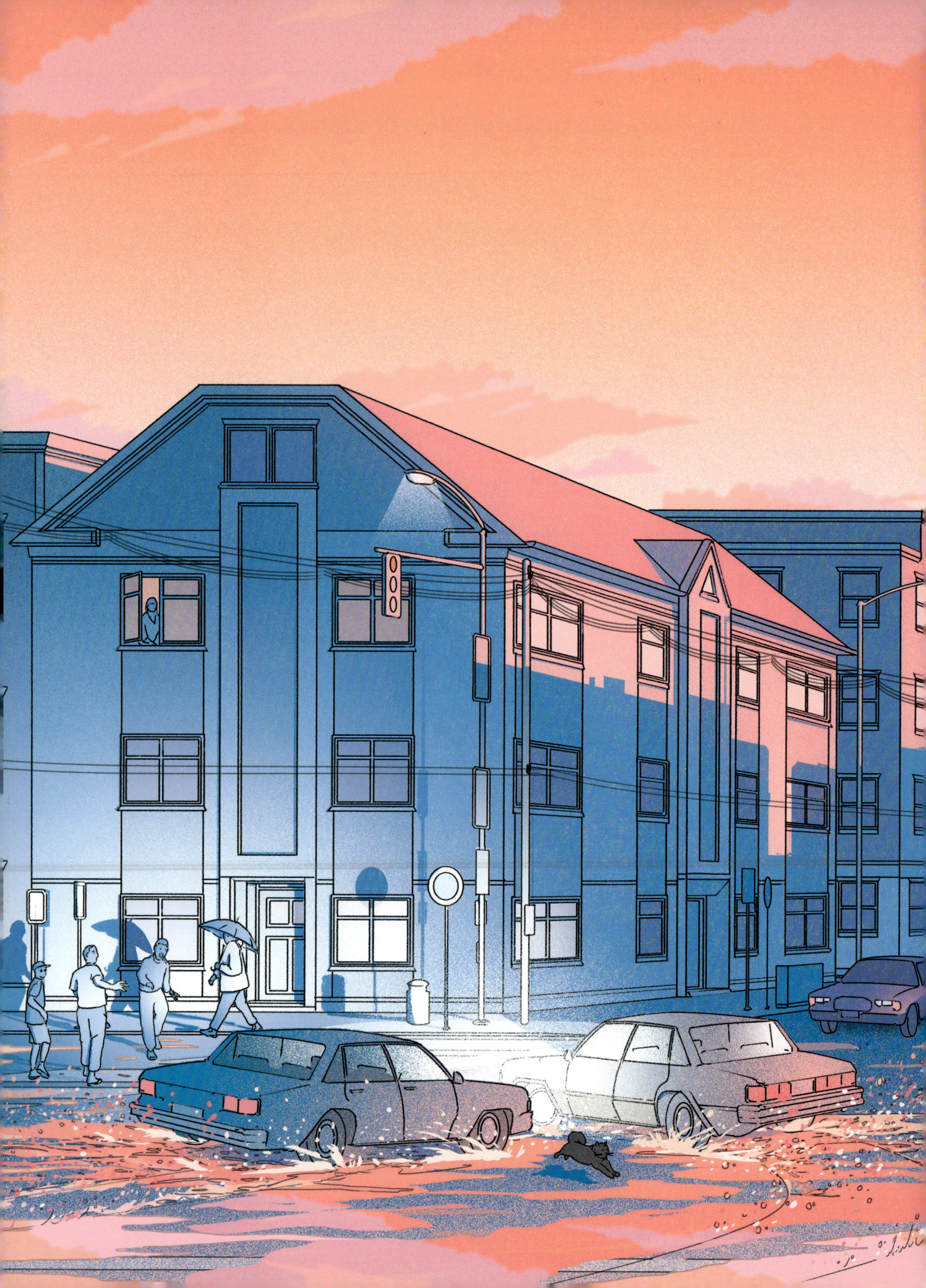

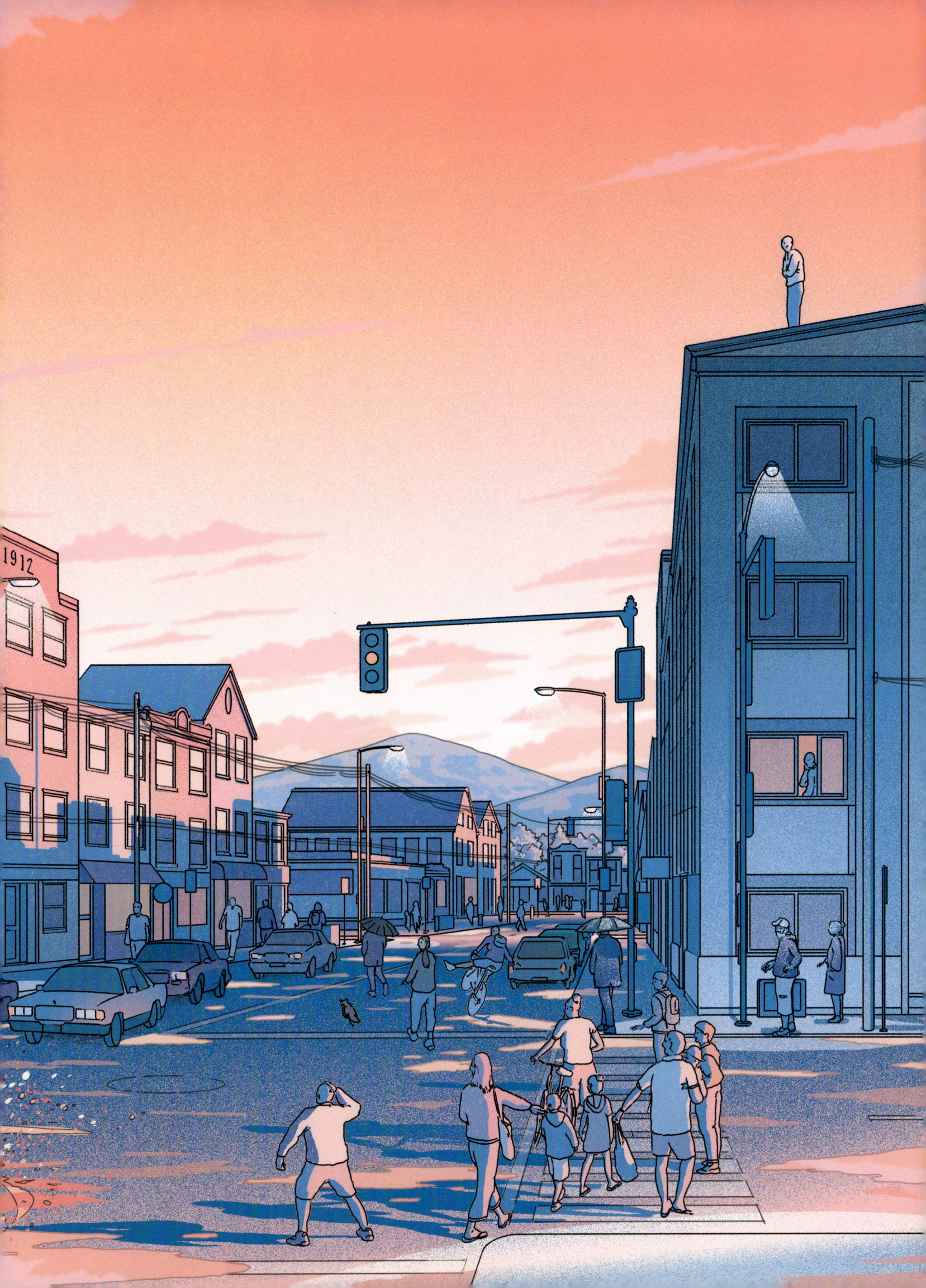

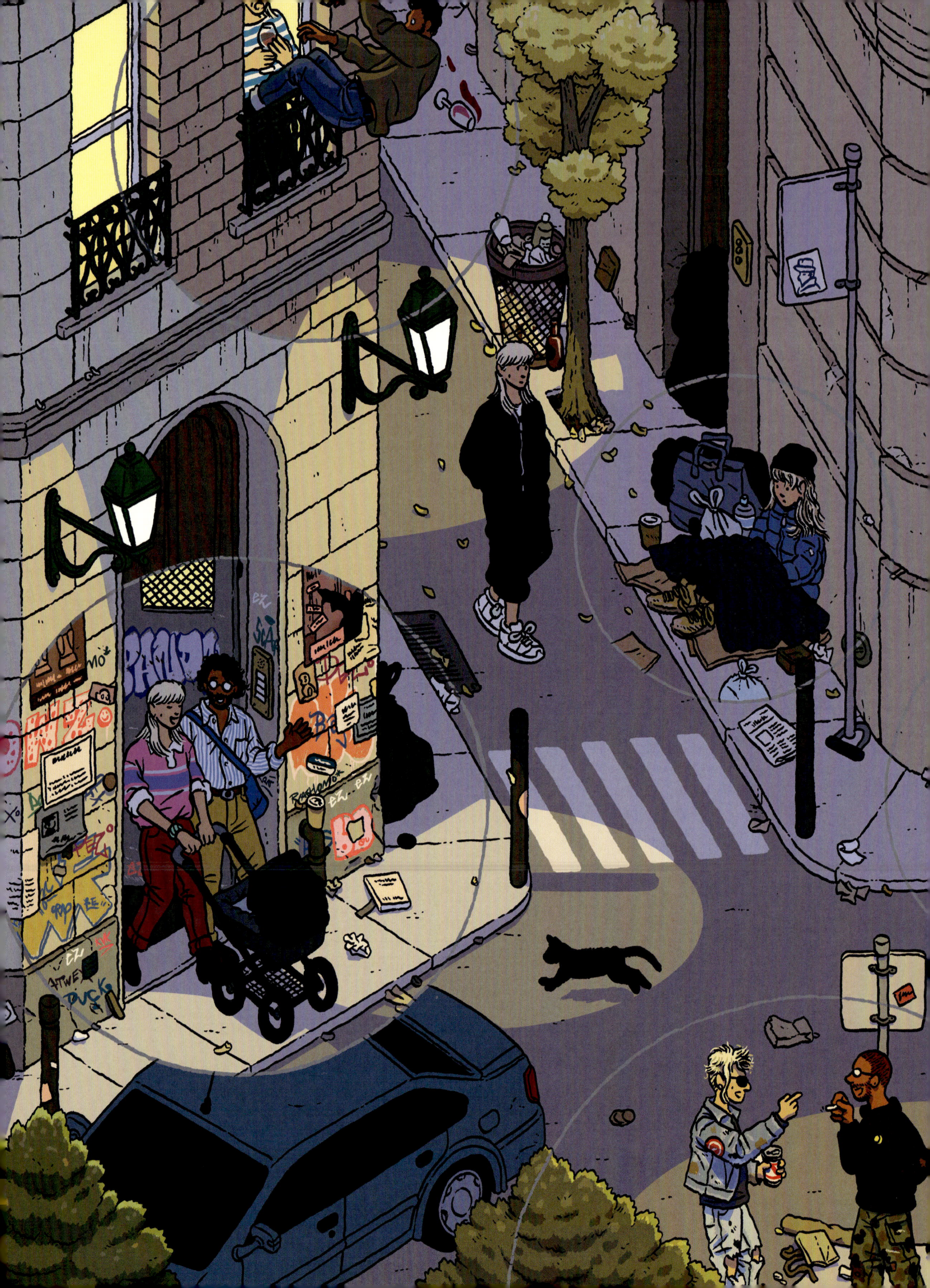

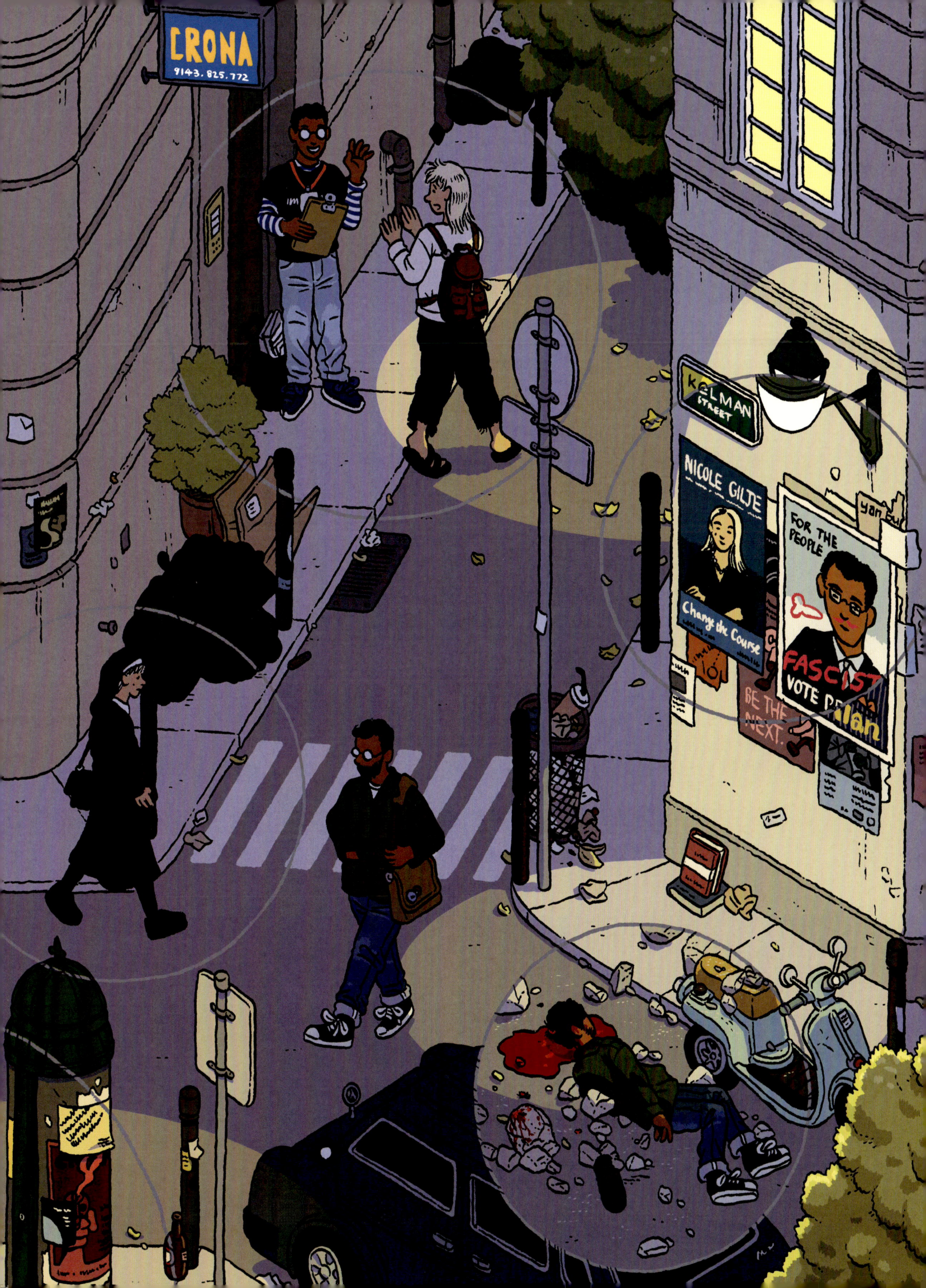

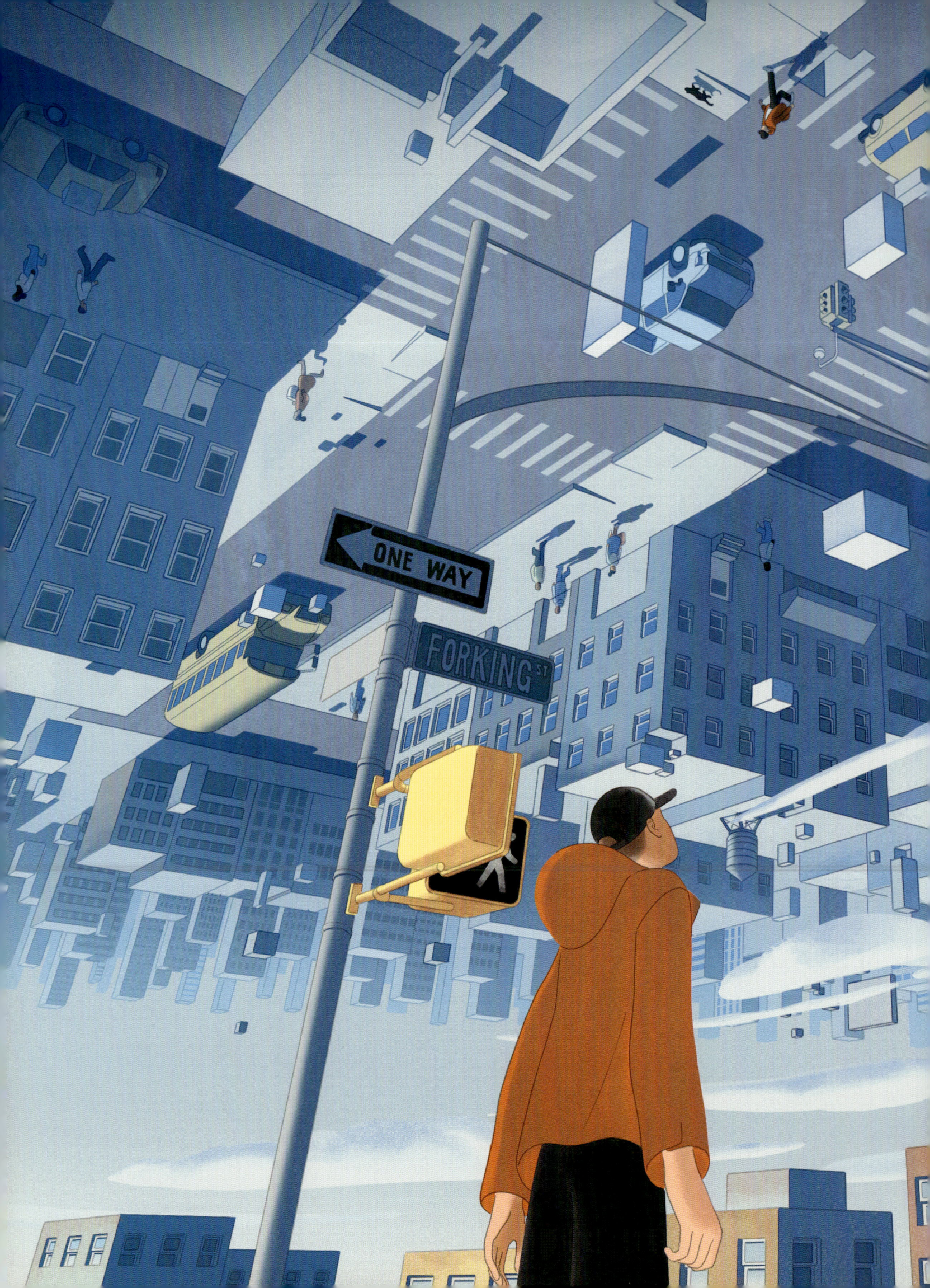

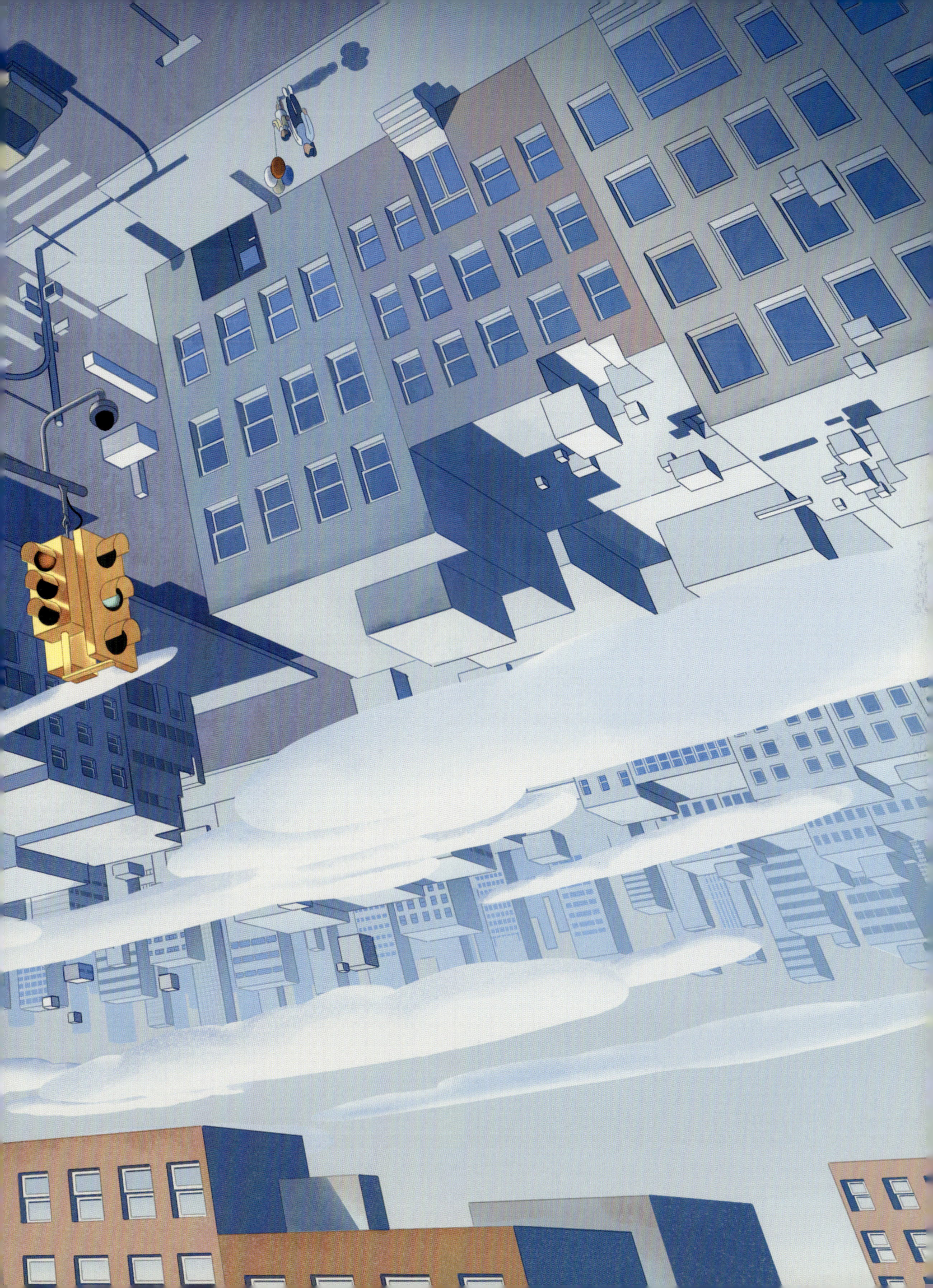

**Portrait of the Alcoholic
Stranded Alone on a Desert Island**
by Kaveh Akbar

I live in the gulf
between what I've been given
and what I've received.

Each morning, I dig into the sand
and bury something I love.
Nothing decomposes.

It might sound ungrateful to say
I expected poetry, but I did—

palm forests and clouds above them
arranged like Dutch still lifes,
musically colored fauna lounging
in perpetual near-smiles.

Instead, these tumors under the surf.

Wildness: to appear
where you are unexpected.

My favorite drugs are far from here.

Our father, who art in Heaven—always
just stepped out, while Earth,
the mother, everywheres around.

It all just means so intensely: bones
on the beach, calls from the bushes,
the scent of edible flowers
floating in from the horizon.

I hold my breath.

The boat I am building
will never be done.

Absence of evidence is not evidence of absence.

Attributed to Carl Sagan or Martin Rees

I want a theory of physics that accounts for the structure of the universe, that clarifies what it is to be an observer in the universe, not a theory that makes the universe depend on me observing it.

Carlo Rovelli, Helgoland

The Many Worlds of Hugh Everett III

by Lizzy Stewart

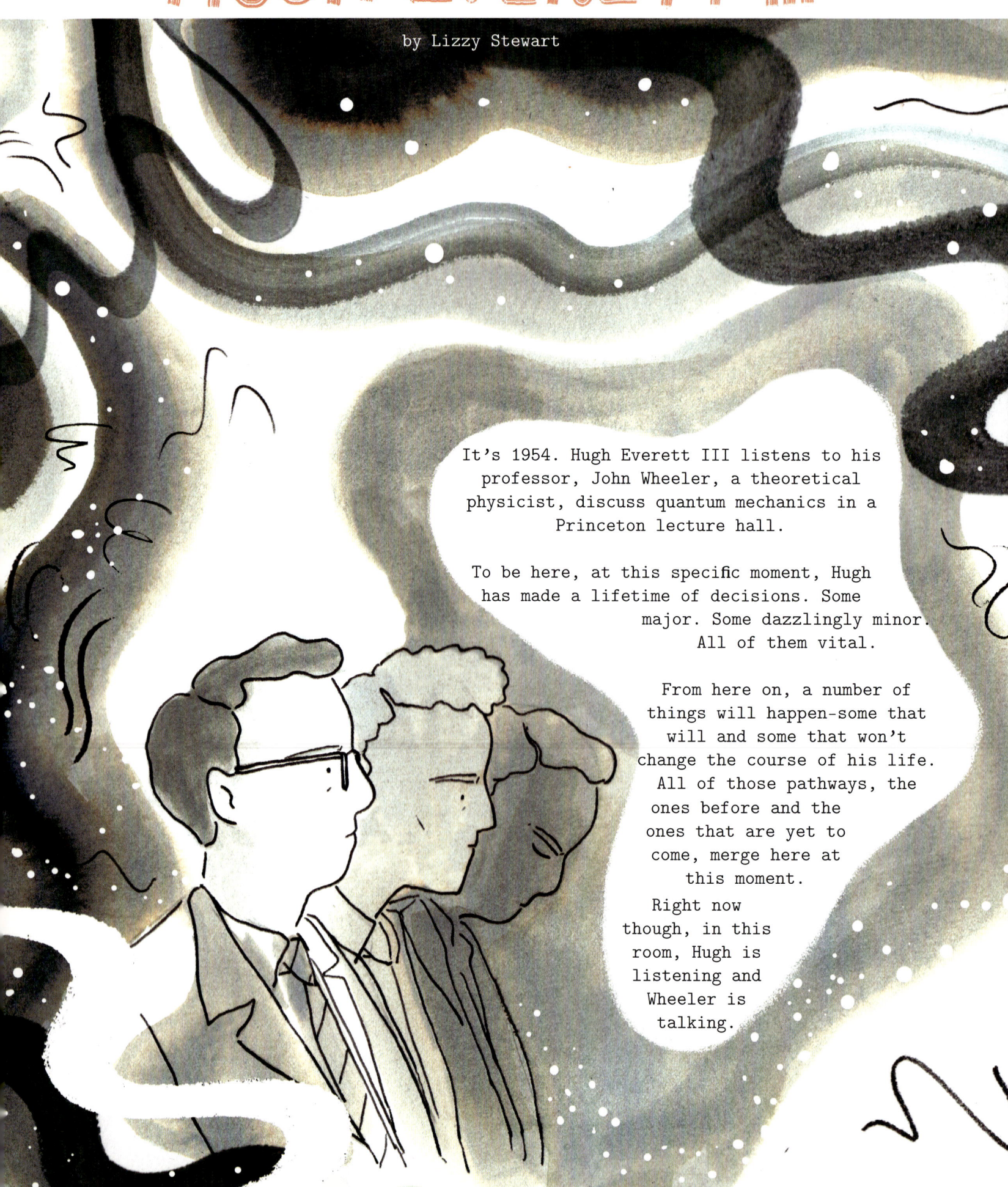

It's 1954. Hugh Everett III listens to his professor, John Wheeler, a theoretical physicist, discuss quantum mechanics in a Princeton lecture hall.

To be here, at this specific moment, Hugh has made a lifetime of decisions. Some major. Some dazzlingly minor. All of them vital.

From here on, a number of things will happen-some that will and some that won't change the course of his life. All of those pathways, the ones before and the ones that are yet to come, merge here at this moment.

Right now though, in this room, Hugh is listening and Wheeler is talking.

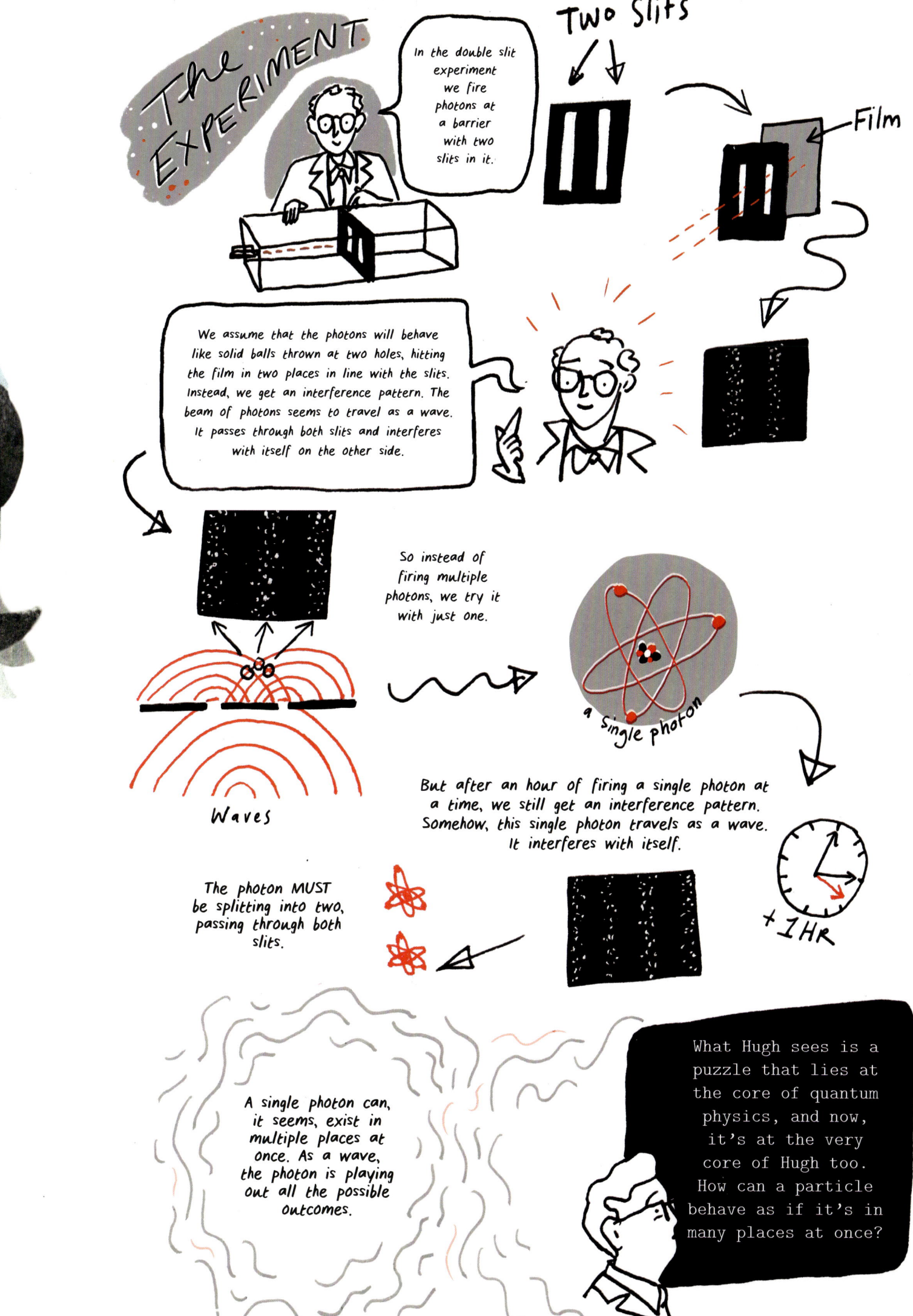

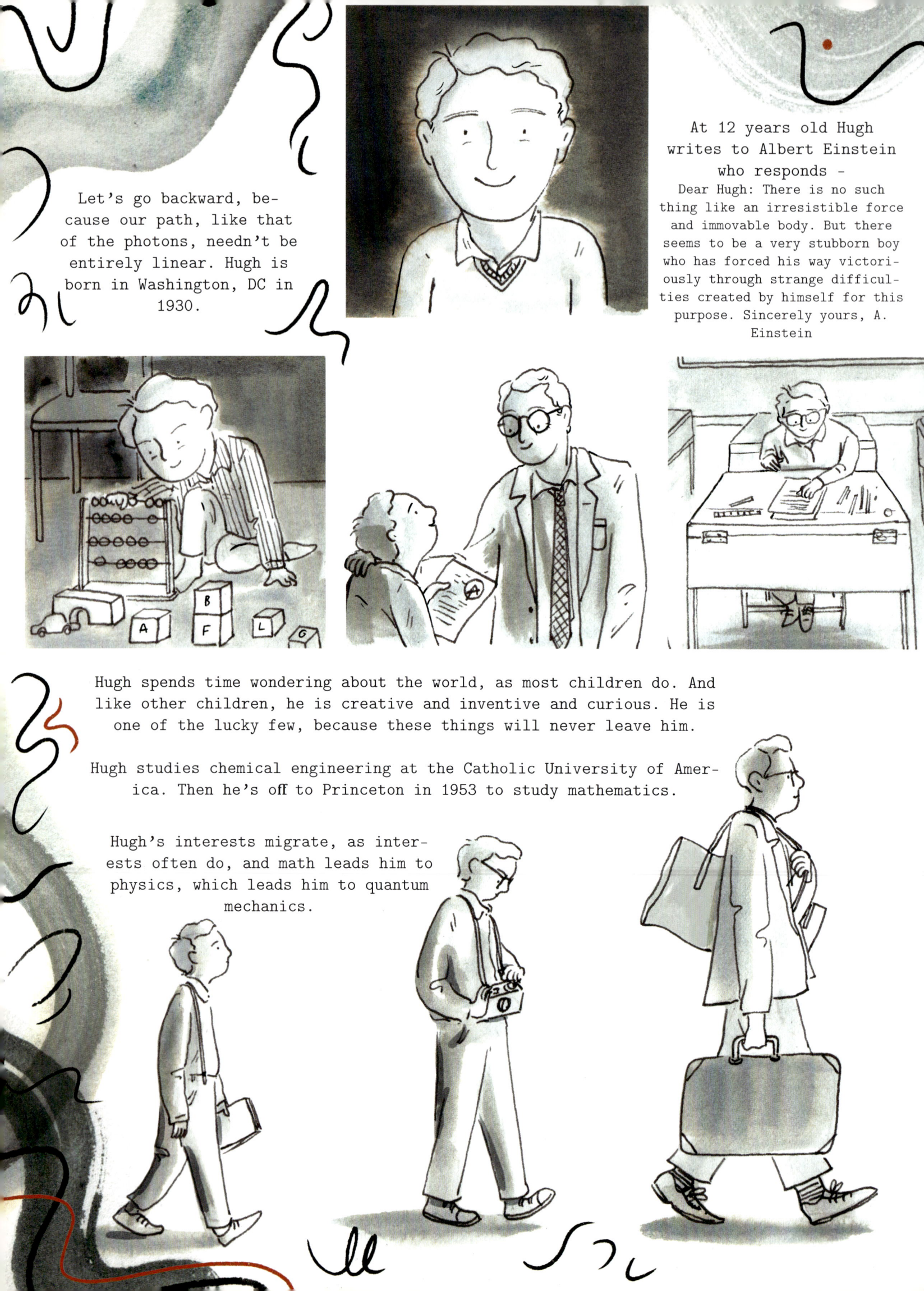

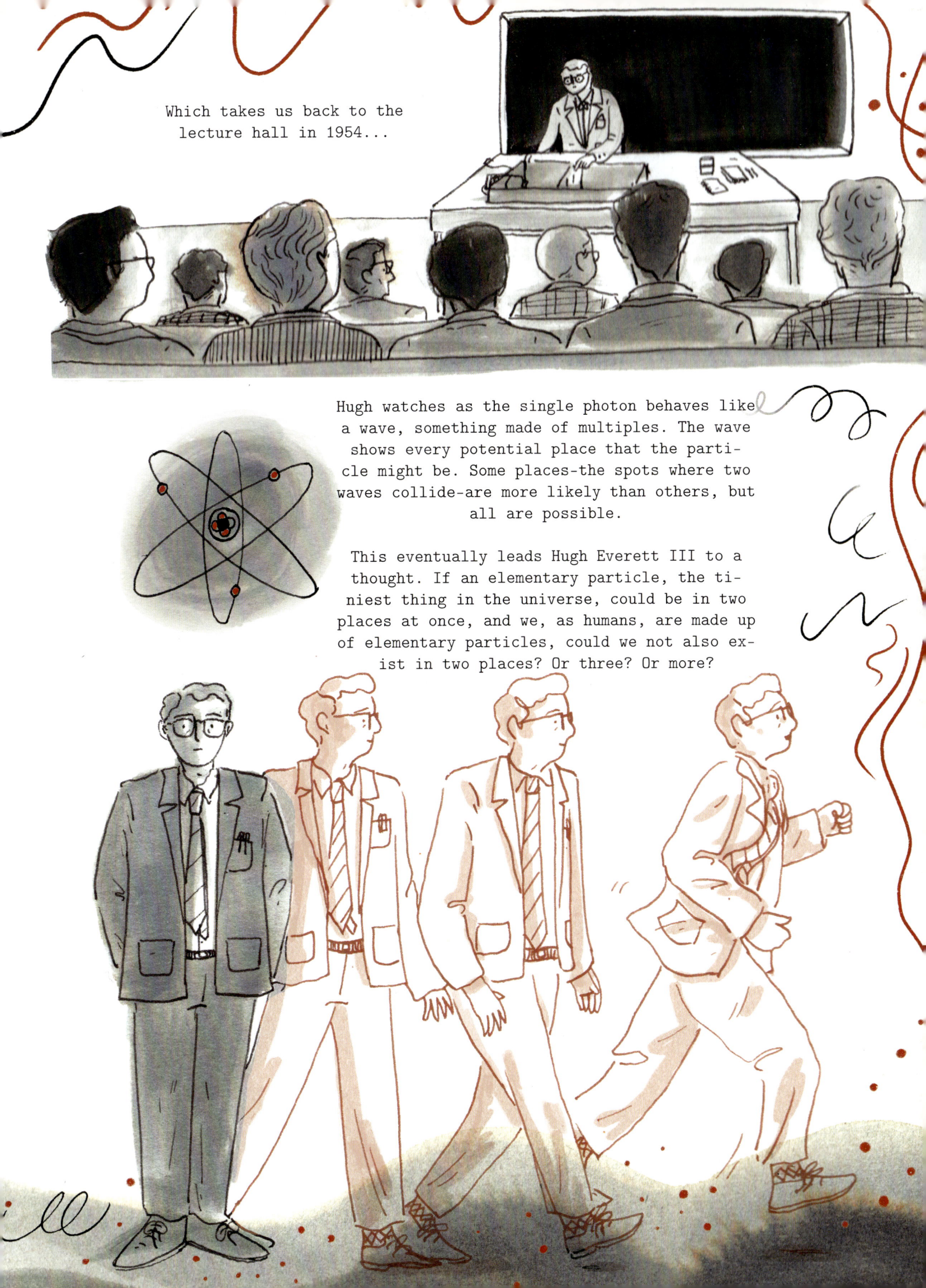

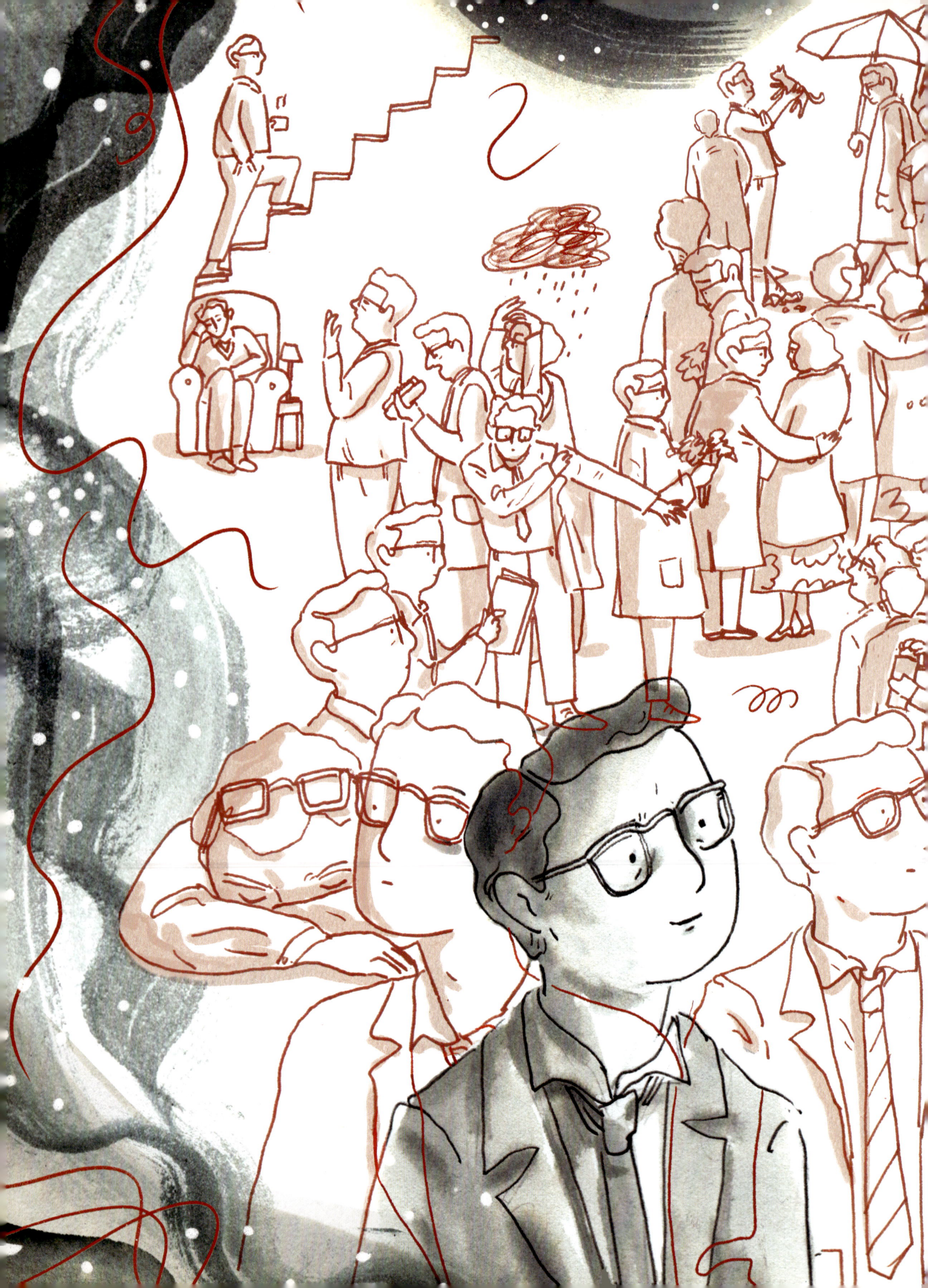

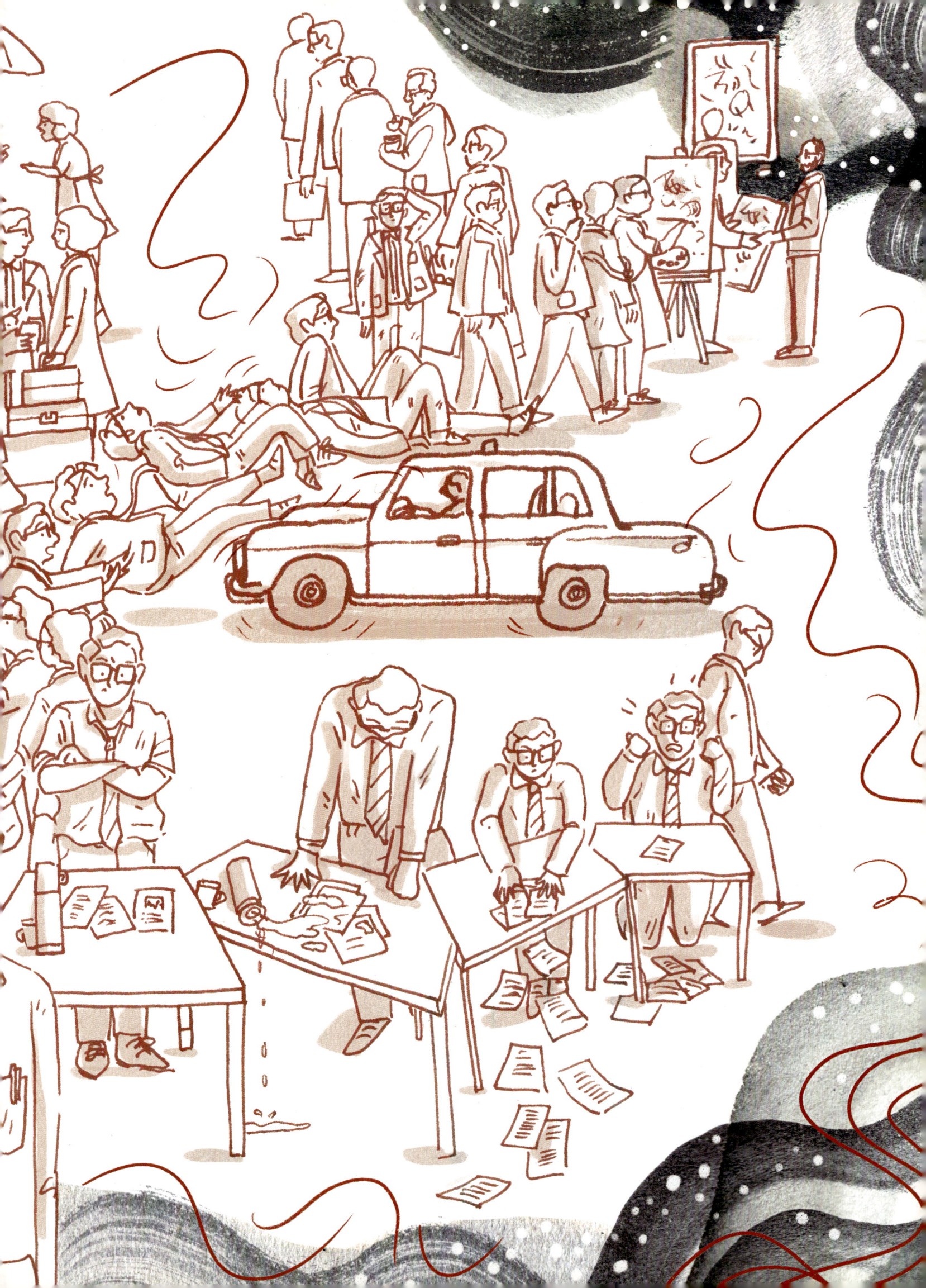

When Hugh writes his dissertation, the many-worlds theory is born. It's a brand-new idea, controversial and, perhaps, eccentric.
His theory posits that every time something happens, the alternate outcomes are played out in a parallel reality.

He suggests that we exist in infinite iterations...

...across infinite worlds where every single possibility is enacted, without us ever seeing it.

Hugh works hard on his theory. In a different universe, he stops writing and starts dog-walking, or becomes fixated on marine life, or dates a woman with a maniacal laugh, or, or, or…

In this universe, though, Hugh Everett III meets Nancy Gore and they get married

He completes his dissertation and starts work at the Pentagon. A few years later he defends his dissertation. Up until now, till Hugh, quantum scientists believed that the atom splits only when un-observed. Hugh says that the atom both splits and doesn't split, it exists in two states, in two parallel worlds.

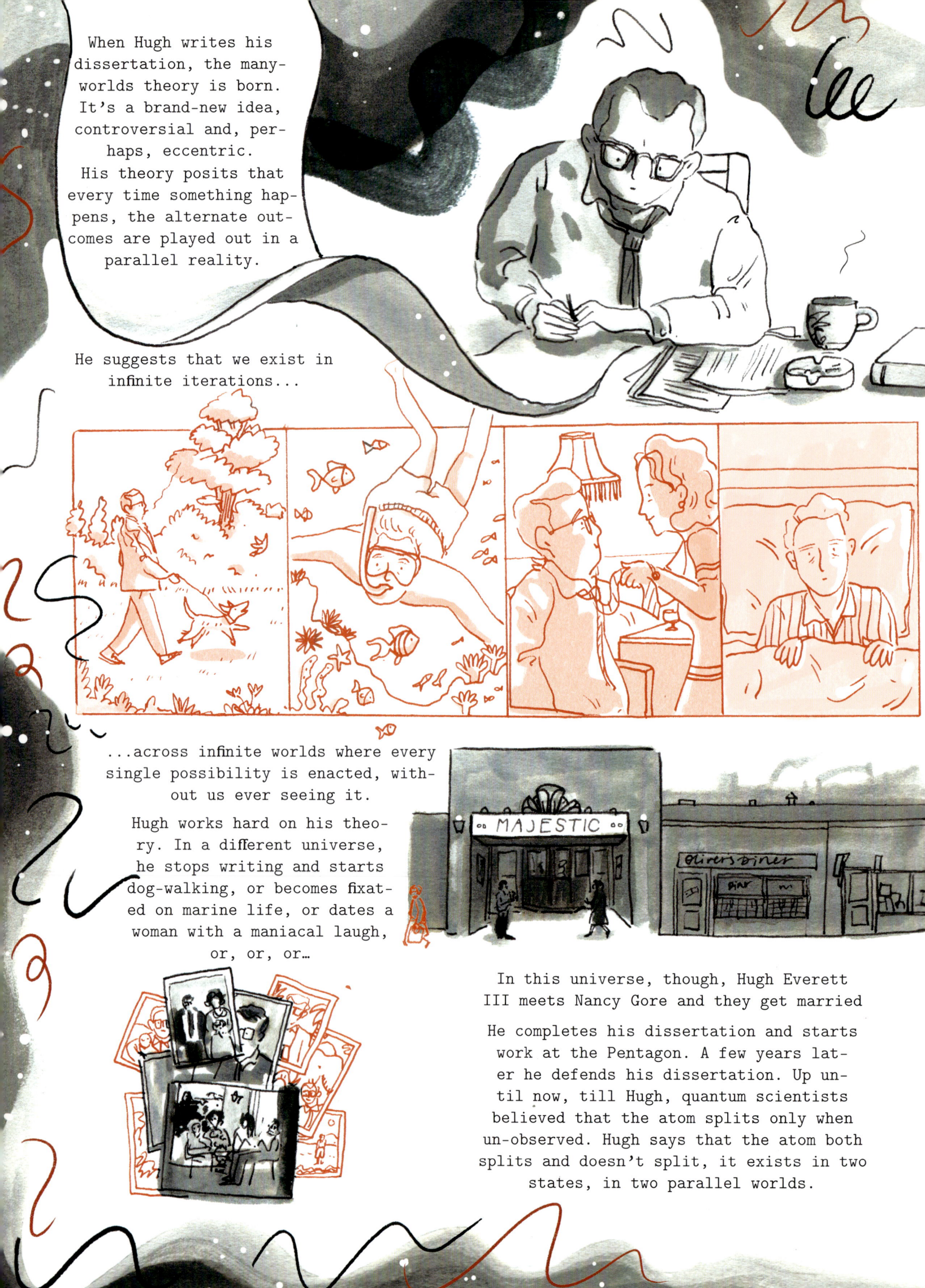

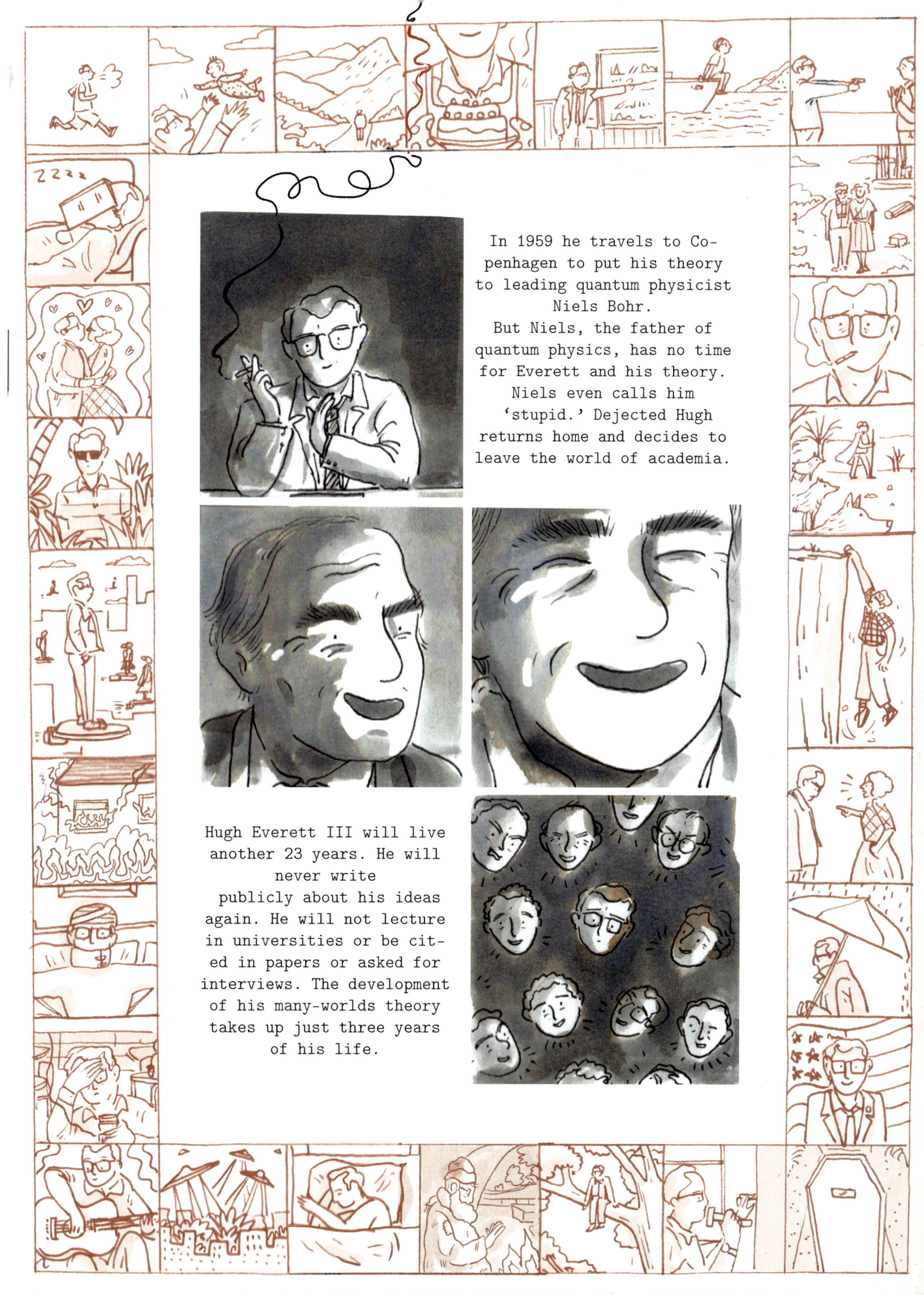

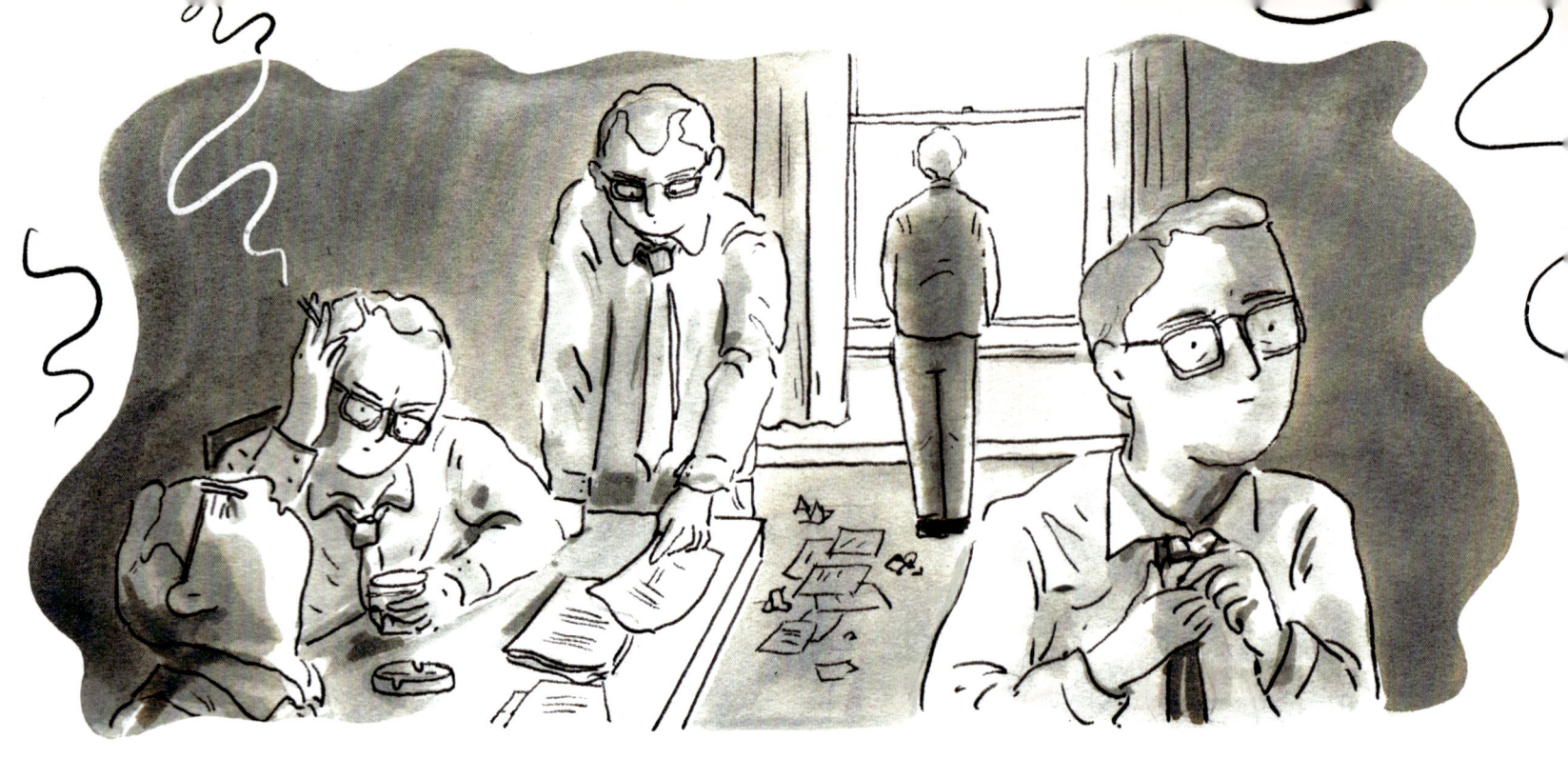

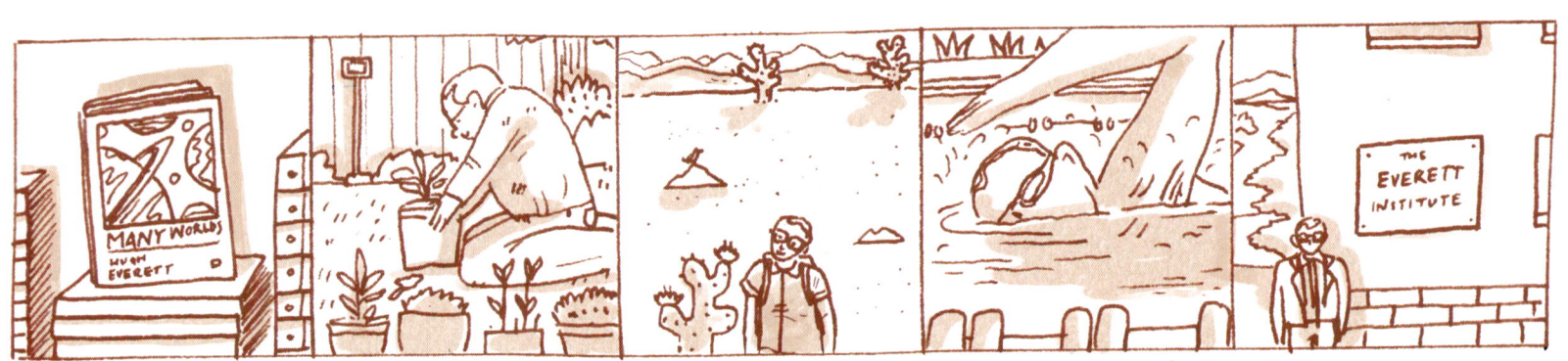

He returns to work. He lends his strange, creative brain to both civil defense and the world of business. He and Nancy have two children, Mark and Elizabeth. He works hard and, quite possibly, smokes too much and drinks too often.

Some of his work, though, is incredibly important. For example, his research helps to enforce limits on the strength of nuclear weapons.

Maybe we can assume that there has been some kind of retreat within Hugh, that Niels Bohr's blow knocked him back in a way that he never recovered from. We don't really know. There is so little in this universe to go on…

In 1970, there's a change. Someone *does* notice Hugh and, in fact, reads his dissertation. Bryce DeWitt, another theoretical physicist, includes Hugh's writing in an anthology of quantum theories. The anthology sells out.

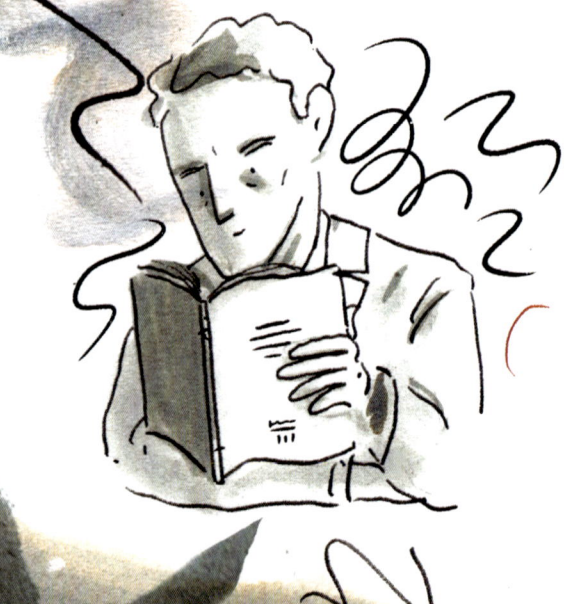

In 1977, Hugh's former professor, John Wheeler, invites him to Texas to talk about the ideas included in the anthology.

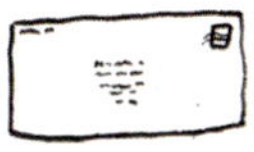

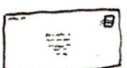

Hugh travels to Texas to speak with John and Bryce....

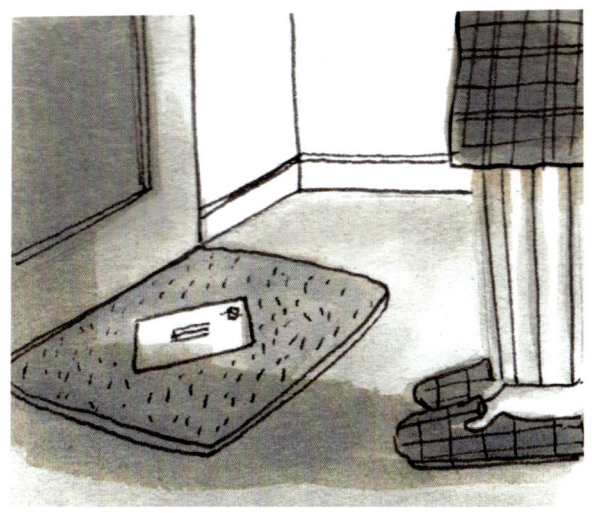

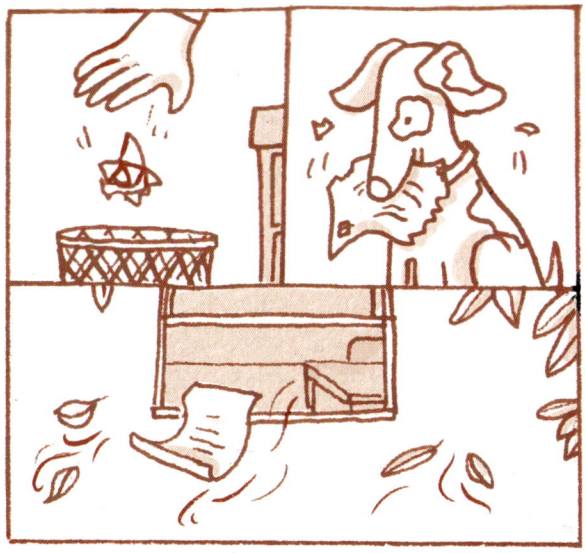

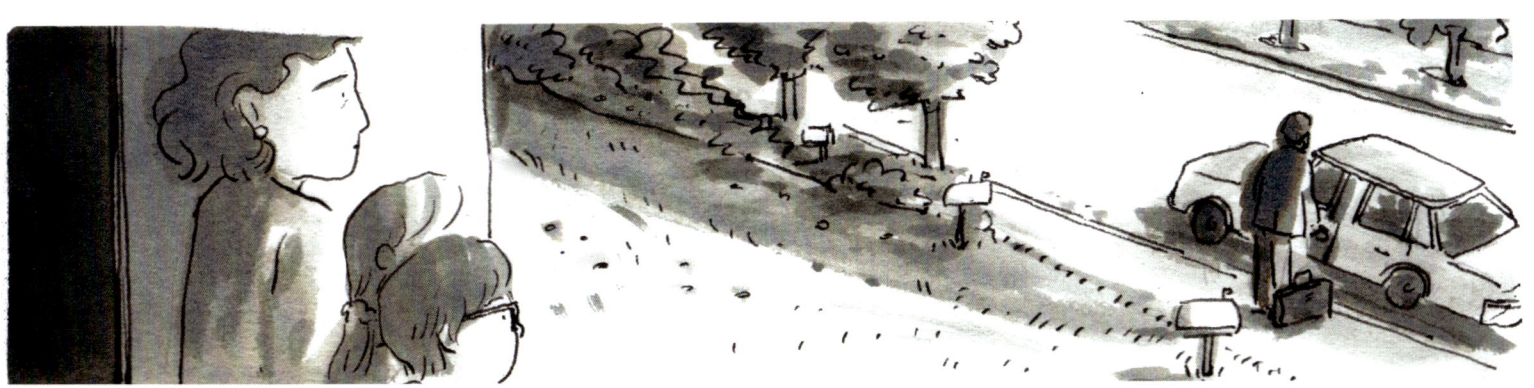

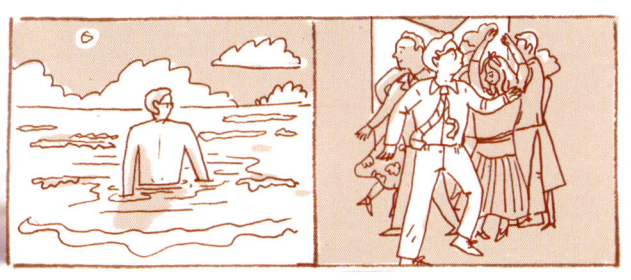

John even suggests that Hugh might make a return to physics.

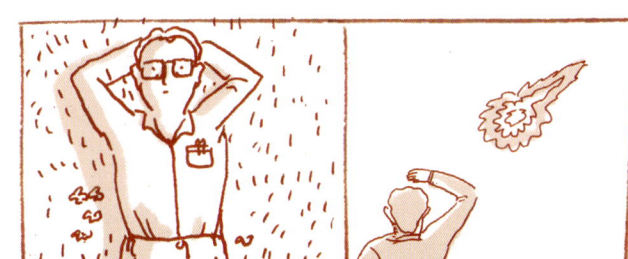

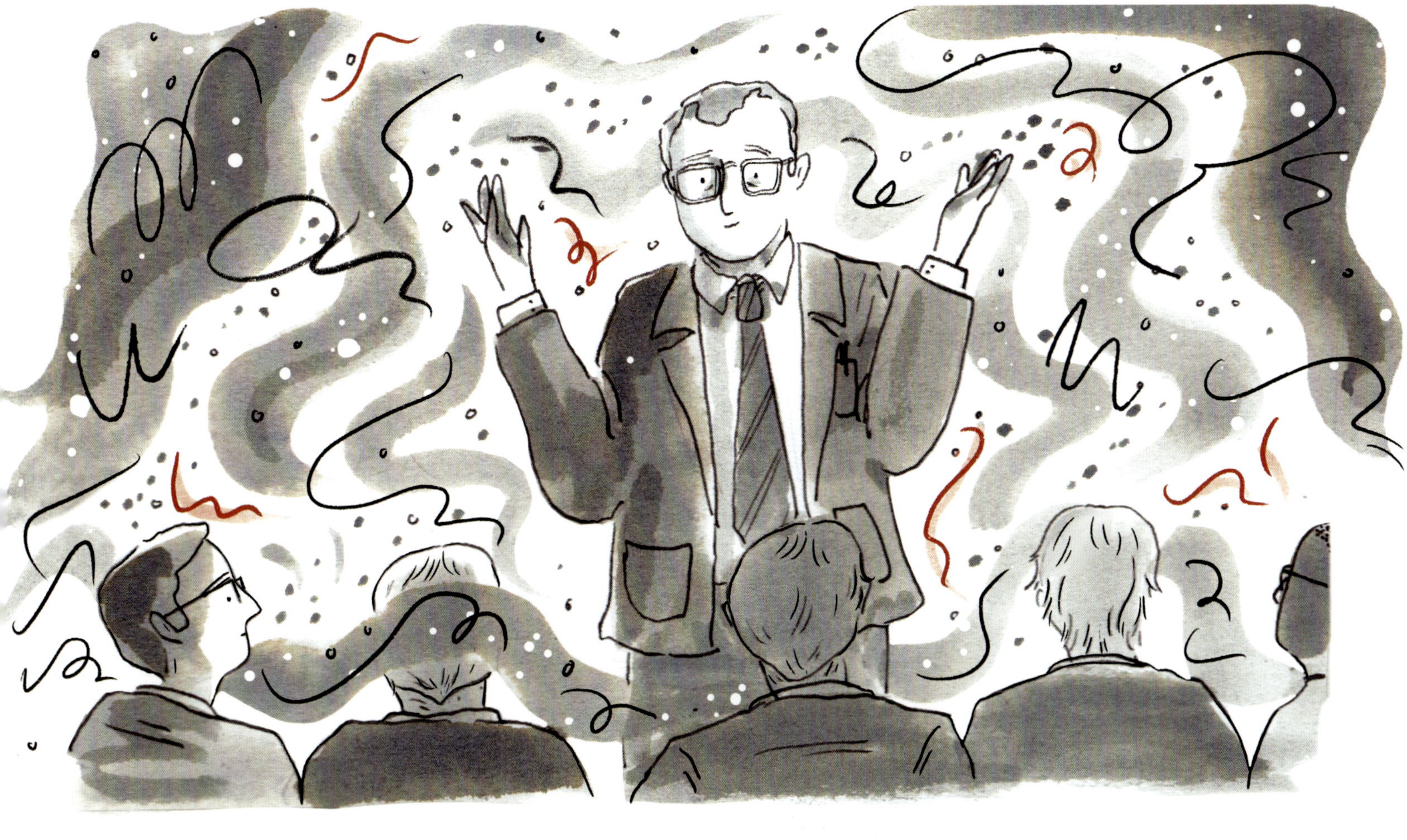

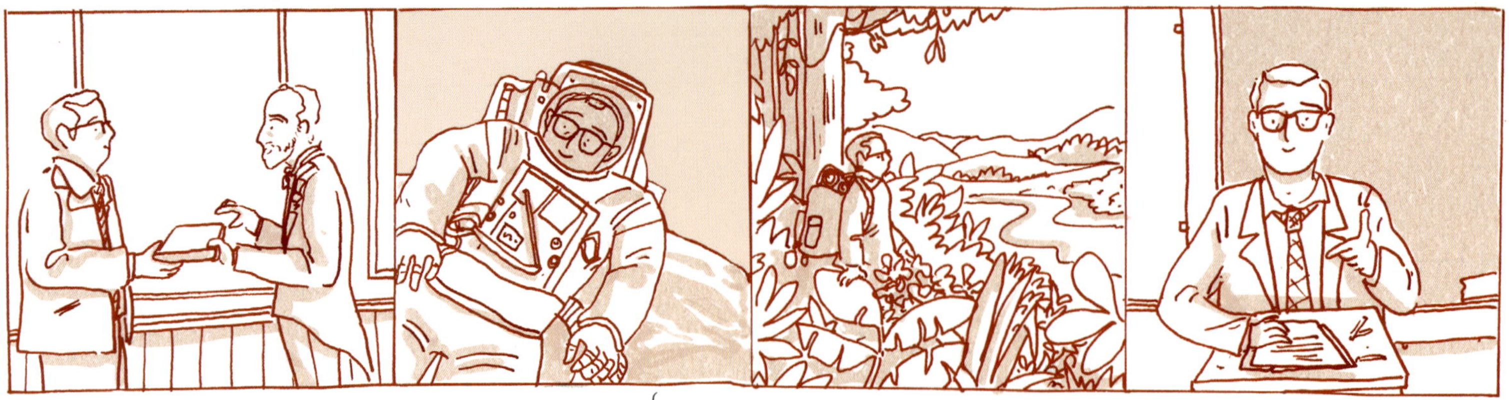

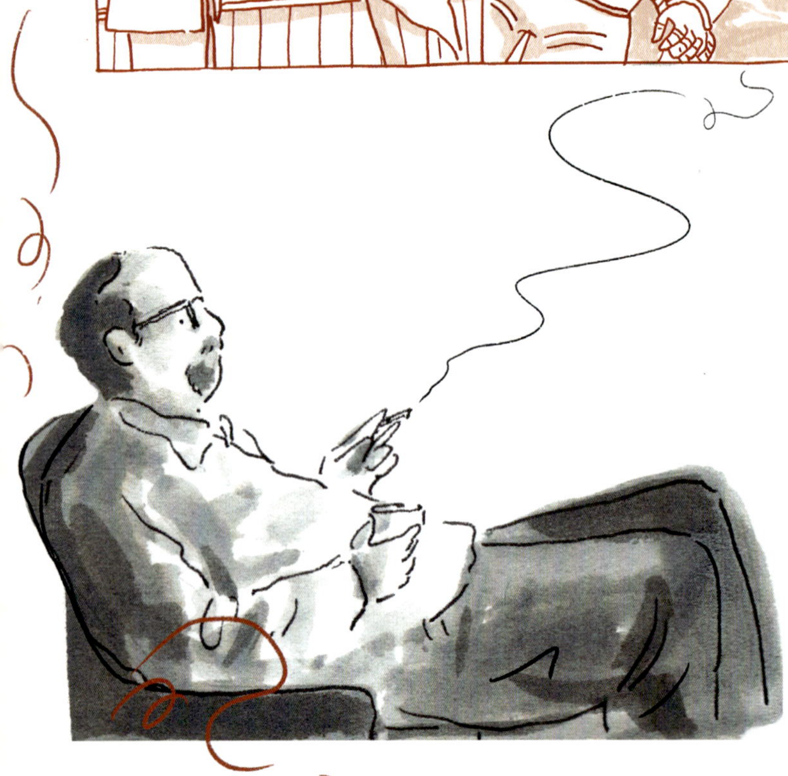

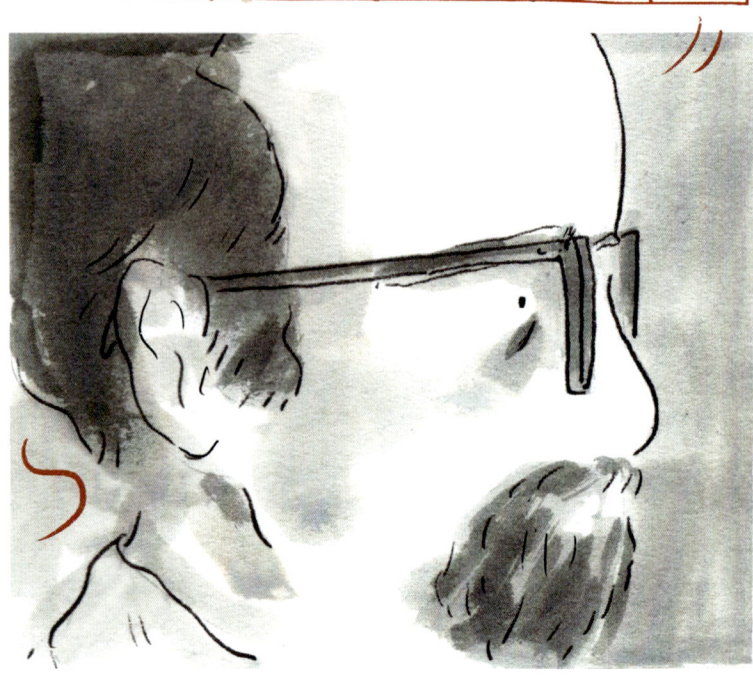

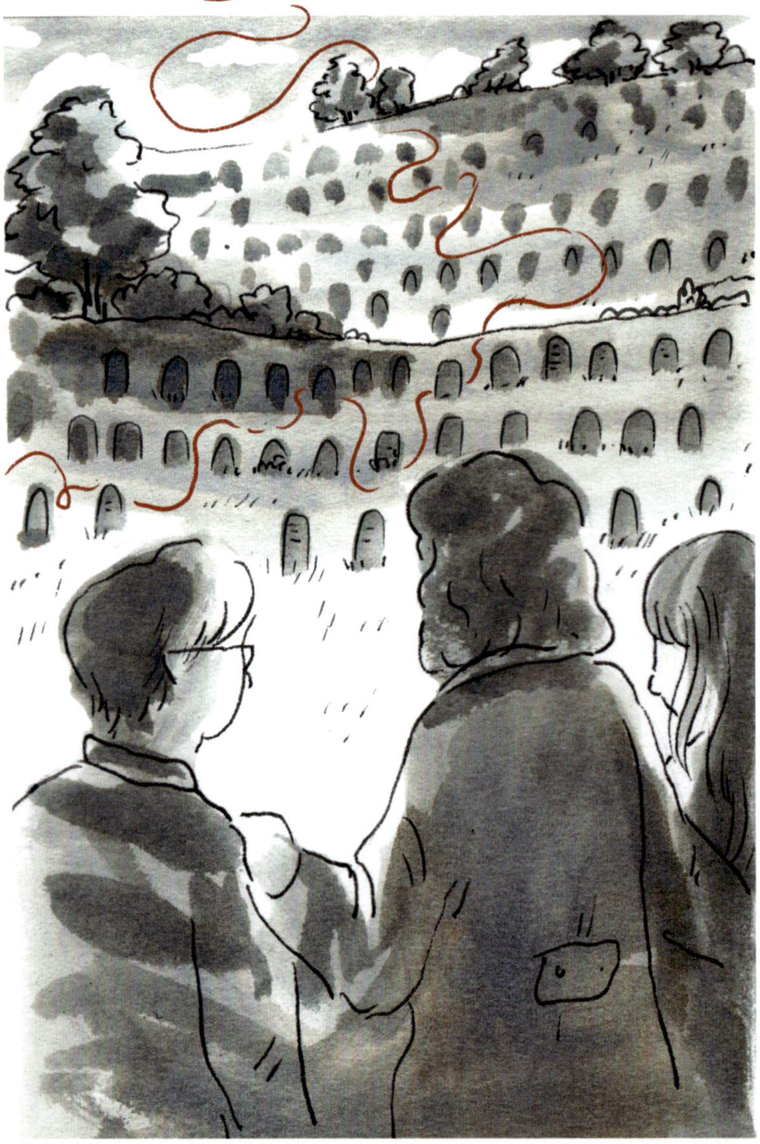

Hugh is enthusiastic, but, in the end, it turns into one of those things. The noises are made, ideas are alluded to, but nothing actually happens. Not in this world anyway.

In the years following his trip to Texas, Hugh is buoyed by the relative success of his visit, but he does not receive much more interest in his many-worlds theory. He lives quietly, something of a mystery to his own family, who struggle to understand the worlds that occupy his thoughts.

On July 19, 1982, Hugh Everett III dies of a sudden heart attack at his home in McLean, Virginia. He is found by his son, Mark.

Hugh Everett III had requested that his remains be thrown in the trash.

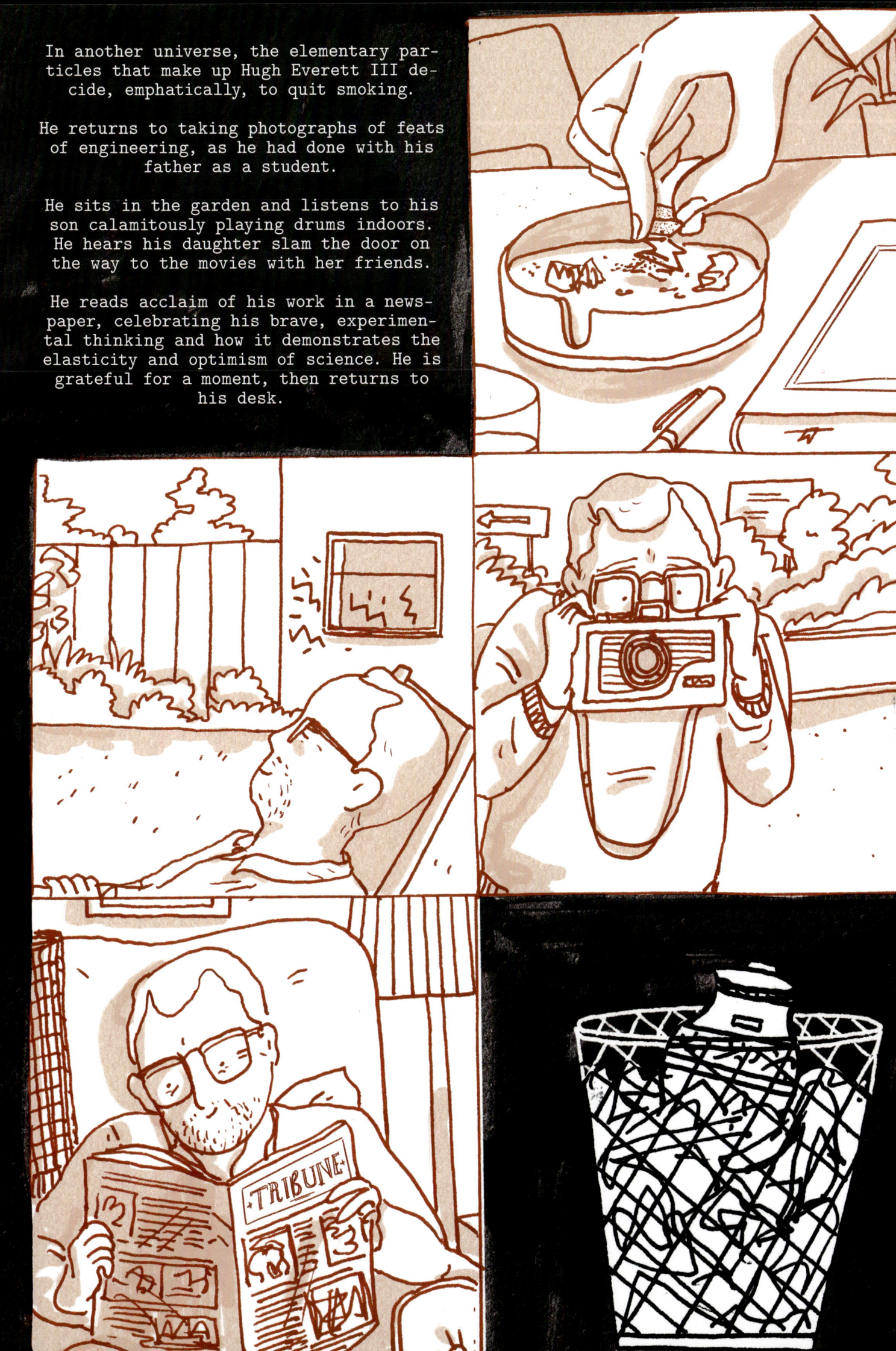

In another universe, the elementary particles that make up Hugh Everett III decide, emphatically, to quit smoking.

He returns to taking photographs of feats of engineering, as he had done with his father as a student.

He sits in the garden and listens to his son calamitously playing drums indoors. He hears his daughter slam the door on the way to the movies with her friends.

He reads acclaim of his work in a newspaper, celebrating his brave, experimental thinking and how it demonstrates the elasticity and optimism of science. He is grateful for a moment, then returns to his desk.

Chapter 2

Quilted

Because light only travels at a finite speed, there's only so much universe we can possibly see or affect or be affected by. Anything beyond our cosmological horizon—the farthest point in every direction from which light has had time to reach us, and our light it since the Big Bang—might as well not exist as far as we're concerned, because there's no accessing it. Still, we'd be crazy not to wonder how much more universe is out there.

From what we can see in our small "patch" of observable universe, it is difficult to know whether the rest of the universe extends infinitely in "flat" space or if space curves until it gradually comes back around in a "closed" universe. What we do know is that, for space to be flat, the density of matter in the universe has to be such that the energy of our universe's expansion balances the energy of gravity's inward pull. And based on calculations and data collected over the last 20 years, we have found that the universe comes within a whisker of this critical density that would indicate that the universe is flat.

Space either curves too gradually for us to notice within our limited field of view or it is truly flat, in which case it is infinite. That would mean that an infinite number of other patches exist beyond our own, forming a multiverse quilt. Each patch contains a finite number of elementary particles—within our cosmological horizon, something like 100,000,000,000,000,000,000, 000,000,000,000,000,000,000,000,000,000,000,000,000,000,000, 000,000,000,000 particles, to be exact—and a finite number of particles can be arranged in only a finite number of ways. With infinite patches in the quilted multiverse, particle arrangements must repeat. In fact, every possible particle arrangement must repeat in countless patches. That means the quilt features no fewer than an infinitude of yous playing out your same story in an infinite number of identical patches, along with innumerable patches containing your close-but-not-exact doppelgängers, patches with people who are extremely similar to you, similar, merely reminiscent, and, of course—if you prefer to think of these—plenty of patches where there's no one like you at all.

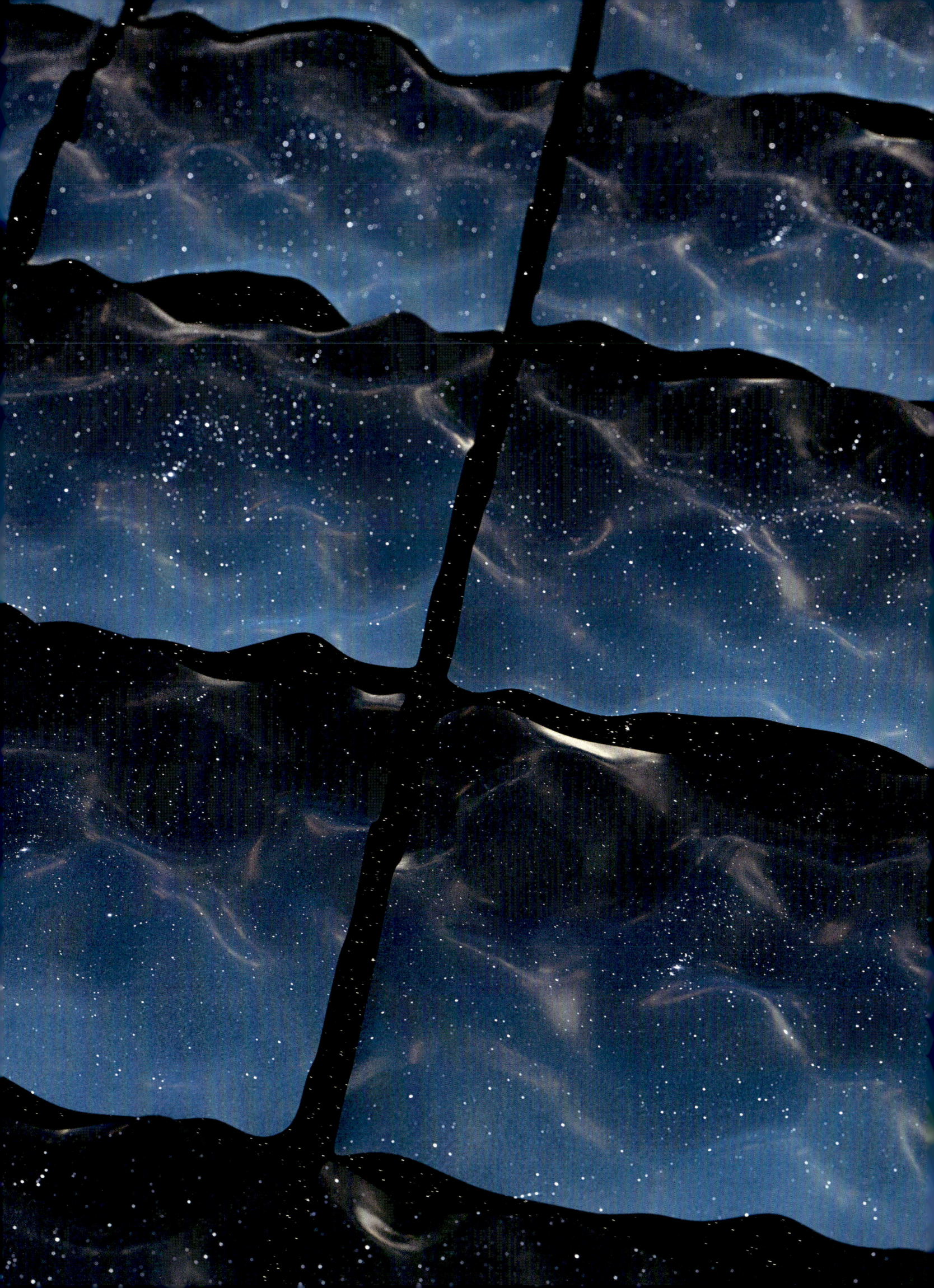

"All philosophy," I told her, "is based on two things only: curiosity and poor eyesight [...] the trouble is, we want to know more than we can see."

Bernard Le Bovier de Fontenelle, *Conversations on the Plurality of Worlds*

72 % of people believe in miracles.

2013 Harris poll

Man is equally incapable of seeing the nothingness from which he emerges and the infinity in which he is engulfed.

Blaise Pascal, "The Misery of Man Without God," Pensees

The passion of surprise and wonder, arising from miracles, being an agreeable emotion, gives a sensible tendency toward the belief of those events from which it is discovered.

David Hume, "Of Miracles"

THE PATCHWORK EFFECT

A roundtable of quilters discuss the quilted multiverse

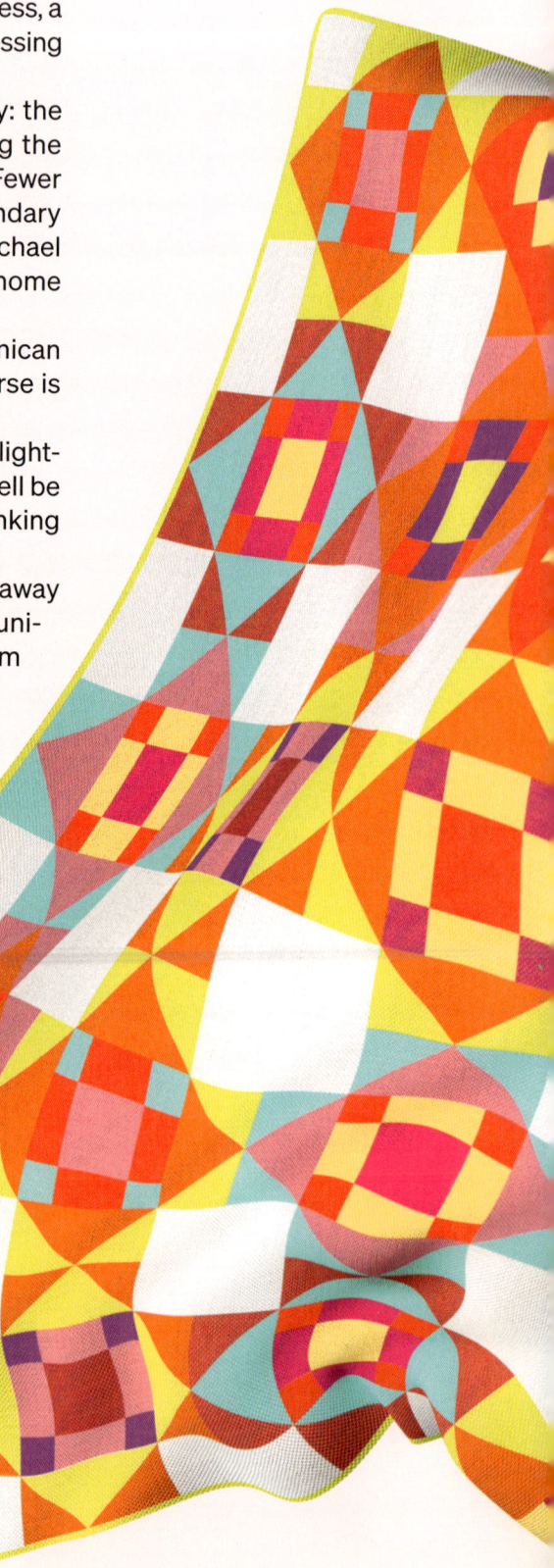

We know our world is round, but we can't see it from where we're standing on the Earth's surface. From a higher vantage point—a window seat on an airplane, for example—one can actually see the curve. The scale is immense, the definition of sublime. But a funny thing happens at an even further vantage point, as in from outer space. There's a cognitive shift in awareness, a shift from feeling the awesome might of a 197-million-square-mile spherical rock, to witnessing that rock as just one minute part of an exponentially more awesome solar system.

Self-described space philosopher Frank White coined the term for this epiphany: the "overview effect." White has spent decades giving words to the sensation of seeing the Earth against the backdrop of the universe, a phenomenon few of us have really felt. Fewer than six hundred humans have ever reached space (an internationally recognized boundary called the Kármán line, roughly sixty-two miles above the Earth). One of the first was Michael Collins, the command module pilot on Apollo 11, who said of his dawning view of our home planet, "The thing that really surprised me was that it projected an air of fragility."

Physicist and cosmologist George F. R. Willis calls this realization a "super-Copernican revolution" in which our planet is not just one among many, but that our entire universe is "insignificant" in the grand, potentially infinite cosmic scale.

About that cosmic scale...the border of our observable universe is roughly 46 billion light-years from where you're standing at present. That unfathomable greatness might as well be made up unless you're a cosmologist, a physicist, or a space philosopher. Still, any thinking person who's lived on the Earth for a number of years has to wonder what lies beyond.

Hypothetically speaking, people living on a planet that's over 46 billion light-years away from Earth—surpassing the border of *our* visibility—would be in their own universe, a universe separate from ours. They can, in theory, see only forty-six billion light-years from where they exist, repeated in all directions into infinity. Can you start to see where I'm going with this? That repetition suggests that infinitely more of these "patches" exist beyond our own, stitching together a multiverse quilt. And this is the framework for a theory of the multiverse called "quilted."

Heady stuff, the kind of thinking that emulates psilocybin's effect on the brain. I wanted to know, can we think of the multiverse as a metaphor for broadening our perceptions, for being less siloed and set in our ways? How does one open up room for universal possibility, beyond the afore-mentioned psychedelics? To answer these questions, I, like so many before me, decided to seek out the wisdom of an older generation with a different lived experience than my own (and, given the topic at hand, I may not be the only "me" asking these questions in the infinite, quilted multiverse).

Turns out, all you have to do is fire up Zoom on a Tuesday morning and gather around a metaphorical table to discuss the concept with three seasoned quilters who live in the Bible Belt. I am profoundly grateful to Becky Scheinert, Luann Johnson, and Rebecca Tait, who dove into the humorous, philosophical, and poetic aspects of this larger-than-life theory of the multiverse.

—Kelsey Keith

Becky Scheinert: I am from a family of quilters. My grandmothers, aunts, mother, my husband's great-grandmother... all quilters. I am located in Guntersville, Alabama, which is actually my hometown—I'm a rerun. I'm president of an art gallery that has a permanent and ongoing quilt exhibit.

Luann has been stuck with me since birth, because she is one of fourteen first cousins on one side of my family. Our mutual grandmother was a quilter.

Luann Johnson: Becky's father was my father's brother. Becky's mother and my mother were both seamstresses, and sometimes I think they tried to outdo each other. I'm in Louisville, Kentucky now.

Becky: Rebecca, we were high school friends. We were in band together, in the science club together, and also in trouble together.

[*laughter*]

Rebecca Tait: I currently live in central Texas, although I am an Alabamian by birth and collegiate education. I come from a family of quilters. My father's mother told me that she made [bed]covers, and I make art. I grew up learning to sew, but I don't think I was even aware or really thought about the fact that my grandmothers were quilters until I became an adult.

Kelsey: I'm curious to know about your own belief systems—threads of faith, morality, religion, whatever you've got.

Becky: I'll take a shot at this. I do believe in an afterlife, but it's not like me eating Reese's Cups. An afterlife does not have to contain the current universe that I'm in. It could be twinkly sparkles in the universe just as easily as it could be a different paradigm of this current world. And in my family, we don't fear death.

Kelsey: How does that affect your day-to-day life, as opposed to someone who's living in fear of death?

Becky: It probably helps me not fear getting older. Just kind of go with it, do the best I can with where I am. Okay. Y'all's turn.

Luann: I take comfort in believing that there is some sort of afterlife. I don't know what it's going to be like, but I do believe it's going to be better than what we have here. And sometimes we cannot imagine anything better than life, but sometimes life is heavy. Some people might think, "How do you know?" Well, it brings me comfort. So doggone it, I'm going to believe it. Life is stressful enough as it is.

Rebecca: I guess that puts me on the hot seat. I was raised in much the same environment that Becky was raised in, but today, my faith is more in science. Science is helping us understand a world that ancient peoples weren't able to understand. When I look at science evolving at the rate at which it's evolving, we are just beginning to understand the world, the galaxy, the universe in which we live.

Who knows? Maybe one day we'll have the ability to understand the multiverse—and explore it in the same way that we're now exploring the physical world that we live in.

Kelsey: It's like five hundred years ago, when people were sailing across the oceans. Setting out into the unknown, dropped into an immense body of water must have been akin to exploring the next universe over. You couldn't know what was on the other side, except through theory.

Rebecca: Though you have to give those people credit for their faith in their gods or God, because oftentimes, that faith was what allowed them to take that step into the unknown. I have faith in the physical laws of the universe. And other life out there, somewhere—I believe that is inevitable.

Kelsey: Physicists studying the multiverse often mention faith, because the theory can't be proven—we can't see, experience, or even really conceptualize it. If there are multiple universes beyond our everyday understanding, what feelings does that raise?

Becky: Whenever I get up in a plane, I'm one of those folks who has to have a window seat, so I can look down and see all the roads with the tiny people scurrying about in their cars and their 18-wheelers. It makes me think about how my perspective is not the only perspective in the world. I would not assume that my view is the one and only. And within that, who knows what all is out there.

Quilted

Luann: When we were young, we had no idea beyond our own reality. That's what we grow up with. But with age, I'm thinking, "Well, why couldn't there be more to what we see? Why not?" I grew up in the Christian tradition, and I have clung to that throughout my life. I have faith in our creator, and if he made this, why couldn't he have made other similar situations—or universes?

Rebecca: I think that if you subscribe to the big bang theory, if you believe that's how our existence began... That initial explosion of matter could have coagulated into all these different universes of which there could be an infinite number. And if you apply constants to those universes, could there be other worlds that have followed the same evolutionary time lines as ours? Sure. I believe that's a possibility.

Luann: There are so many things that cannot be proven, but also cannot be disproven.

Kelsey: Anything that comes to mind?

Rebecca: I believe in karma. What you sow comes back to you. Live your life in a way where you treat others the way you want to be treated. That's the way that I was raised. That's the way I raised my daughters. That's what I taught my students. Karma will either reward you or bite you, depending on what you put into it.

Becky: Some people do get their just deserts.

Luann: I find myself getting angry when they don't. We're not the ones to create this karma.

Rebecca: That Martin Luther King, Jr. quote about the arc of the universe bending toward justice; I mean, does it always get there? I just try to tell myself that it may not happen at the moment when I can see it, but the world is bending in that direction, toward what's deserved. And what you've left behind, good or evil, may perpetuate itself after you're gone.

Kelsey: Switching gears for a minute—what does quilting perpetuate for you? What's so appealing?

Luann: I'm getting rid of the humdrum things that don't add any value to my life. But quilts are my happy place.

Rebecca: I quilt because it makes me happy. When I'm gone, my daughters and grandchildren can take them all out in the backyard, put them in a pile, and set fire to them. And I won't know, because I'll be dead. But in the meantime, they make me happy.

Becky: There's something about quilts that's just a major human connection. They bring a warm feeling even if you're not cold.

Kelsey: How do you all relate the practice of quilting to the multiverse?

Rebecca: When I look at a quilt, the first thing I always look for is the square. What's the square? When you first look at any patch, it's like chaos. And then my mind starts trying to make order out of chaos, I guess, kind of like the universe.

Becky: My grandmother would collect scraps. She had ten kids that grew up and got married and had kids. On that side, there were 24 first cousins. My grandmother would collect scraps from each of those ten families. So we'd all hold corners of her newest quilt and say, "Okay. There's my Easter dress, there's my this." Each of those interpretations was its own little universe, and you'd have multiple generations of fabrics represented in a single quilt.

Kelsey: So the quilt sort of exists in multiple dimensions. The thing itself is one visible object to everyone in the group, but it holds distinct memories and associations for each person within that group.

Becky: What Luann would see in a quilt would be different from what I would see in the quilt, just because of the pieces that we personally relate to.

Luann: I think of *Alice in Wonderland*, when she went down the rabbit hole. Each little scrap [in the quilt] would be like its own hole, with its own world underneath. So kind of a multiverse, with all these simultaneous stories taking place.

Kelsey: As people who are constantly piecing things together—do you spot patterns in everyday life, or does quilting make you think about patterns differently?

Rebecca: I love to make log cabin quilts [a traditional design employing a repeated, nesting motif of single blocks]. I know that sounds pretty simplistic, but to me, they're a metaphor to represent your family, your extended family, your community, and it all starts in the middle square. I use the same palette of fabrics, the same size for my seams—these are my constants that hold it together. But the end result is always something different.

I like doing things that make order out of chaos. I like things that are mathematically coherent.

Kelsey: So in your approach to log cabin, or any pattern, really—you begin them with the same rigor, but they each expand into something different. How does that relate to the multiverse?

Rebecca: The quilted multiverse theory supposes that although all of those different universes were formed in the big bang, all of the constants—like the speed of light, for example—stay the same within each one of those universes. I'd never thought about this, but to me, there you go. What's important to me with quilting is to have the constants. The precision of it, the mathematical unity within the quilt…those are as important to me as whatever pattern I'm working on.

Luann: I see the multiverse more as a crazy quilt, because with multiple universes, why should every universe be exactly the same? In crazy quilts, you can use all kinds of different fabrics, all kinds of stitches, all kinds of seam allowances, whatever you want to do with buttons and lace and embroidery. I think multiverses probably have a little of everything.

73

Hymn to Time
by Ursula K. Le Guin

Time says "Let there be"
every moment and instantly
there is space and the radiance
of each bright galaxy.

And eyes beholding radiance.
And the gnats' flickering dance.
And the seas' expanse.
And death, and chance.

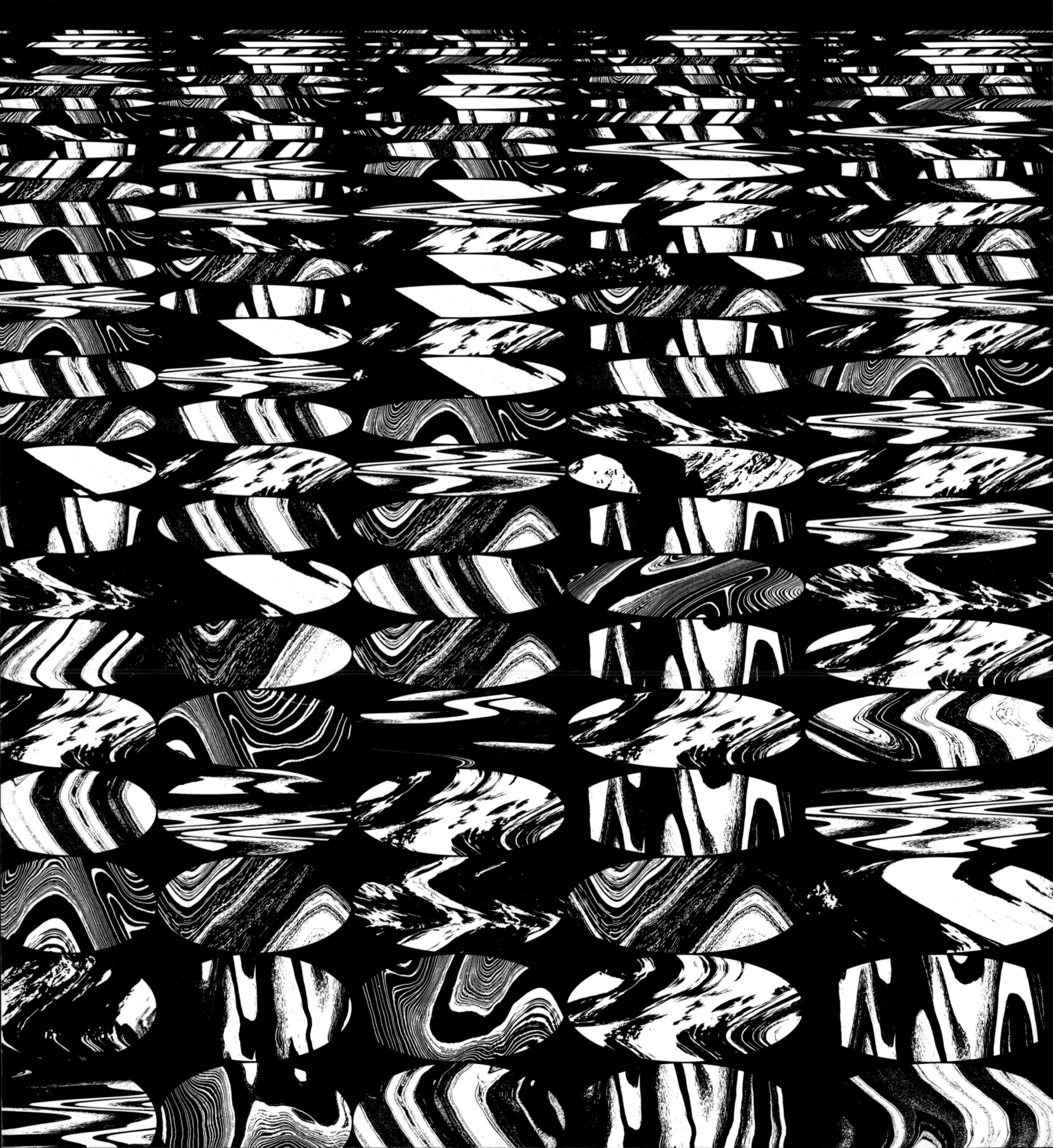

Rebecca: I love to make log cabin quilts [a traditional design employing a repeated, nesting motif of single blocks]. I know that sounds pretty simplistic, but to me, they're a metaphor to represent your family, your extended family, your community, and it all starts in the middle square. I use the same palette of fabrics, the same size for my seams—these are my constants that hold it together. But the end result is always something different.

I like doing things that make order out of chaos. I like things that are mathematically coherent.

Kelsey: So in your approach to log cabin, or any pattern, really—you begin them with the same rigor, but they each expand into something different. How does that relate to the multiverse?

Rebecca: The quilted multiverse theory supposes that although all of those different universes were formed in the big bang, all of the constants—like the speed of light, for example—stay the same within each one of those universes. I'd never thought about this, but to me, there you go. What's important to me with quilting is to have the constants. The precision of it, the mathematical unity within the quilt...those are as important to me as whatever pattern I'm working on.

Luann: I see the multiverse more as a crazy quilt, because with multiple universes, why should every universe be exactly the same? In crazy quilts, you can use all kinds of different fabrics, all kinds of stitches, all kinds of seam allowances, whatever you want to do with buttons and lace and embroidery. I think multiverses probably have a little of everything.

Kelsey: Even in a designed universe, there are bound to be moments of chaos, right?

Rebecca: That's the random nature of it.

Luann: I've been in lots of different quilt guilds, but two years ago, I joined a local Modern Quilt Guild and have learned to be more comfortable doing improvisational quilting, because the people who lead the group say, "There are no rules; you don't have to buy into everything that is in the rule book." Some of those quilts have been such an inspiration. And I don't know—I think the universe might just be an improvisation.

References

Betz, Eric. "Where Is the Edge of the Universe?" *Discover Magazine*. November 6, 2020. https://www.discovermagazine.com/the-sciences/where-is-the-edge-of-the-universe.

Chang, Kenneth. "For Apollo 11 He Wasn't on the Moon. But His Coffee Was Warm." *New York Times*. July 16, 2019. https://www.nytimes.com/2019/07/16/science/michael-collins-apollo-11.html.

Ellis, George F. R. "Why the Multiverse May Be the Most Dangerous Idea in Physics." *Scientific American*. August 2014. https://www.scientificamerican.com/article/why-the-multiverse-may-be-the-most-dangerous-idea-in-physics/.

"International Space Station Facts and Figures." *NASA*. November 4, 2021. https://www.nasa.gov/feature/facts-and-figures.

King, Dr. Martin Luther, Jr. "Remaining Awake Through a Great Revolution."

"The Overview Effect." *NASA*. August 30, 2019. https://www.nasa.gov/johnson/HWHAP/the-overview-effect.

Hymn to Time
by Ursula K. Le Guin

Time says "Let there be"
every moment and instantly
there is space and the radiance
of each bright galaxy.

And eyes beholding radiance.
And the gnats' flickering dance.
And the seas' expanse.
And death, and chance.

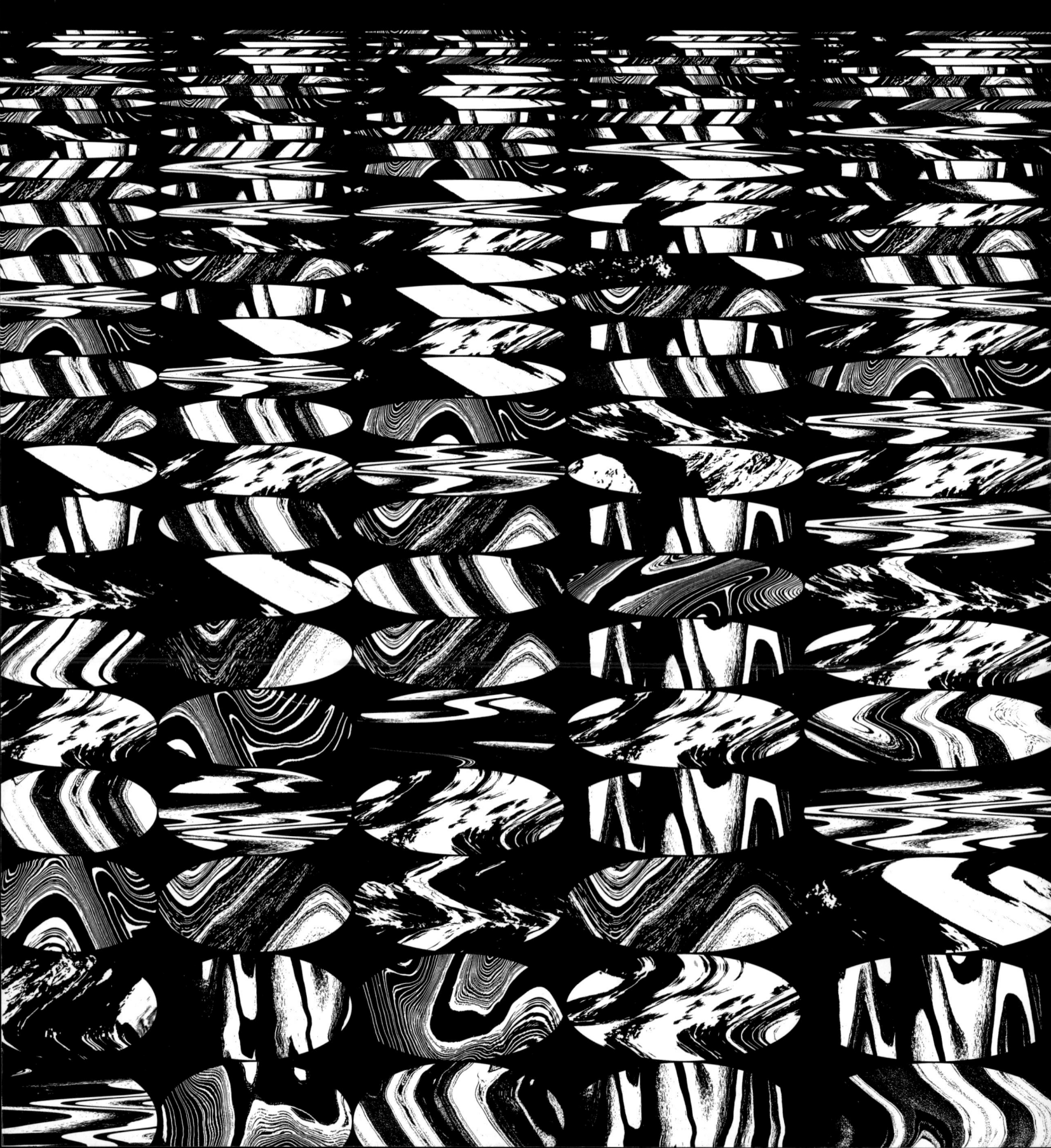

Time makes room
for going and coming home
and in time's womb
begins all ending.

Time is being and being
time, it is all one thing,
the shining, the seeing,
the dark abounding.

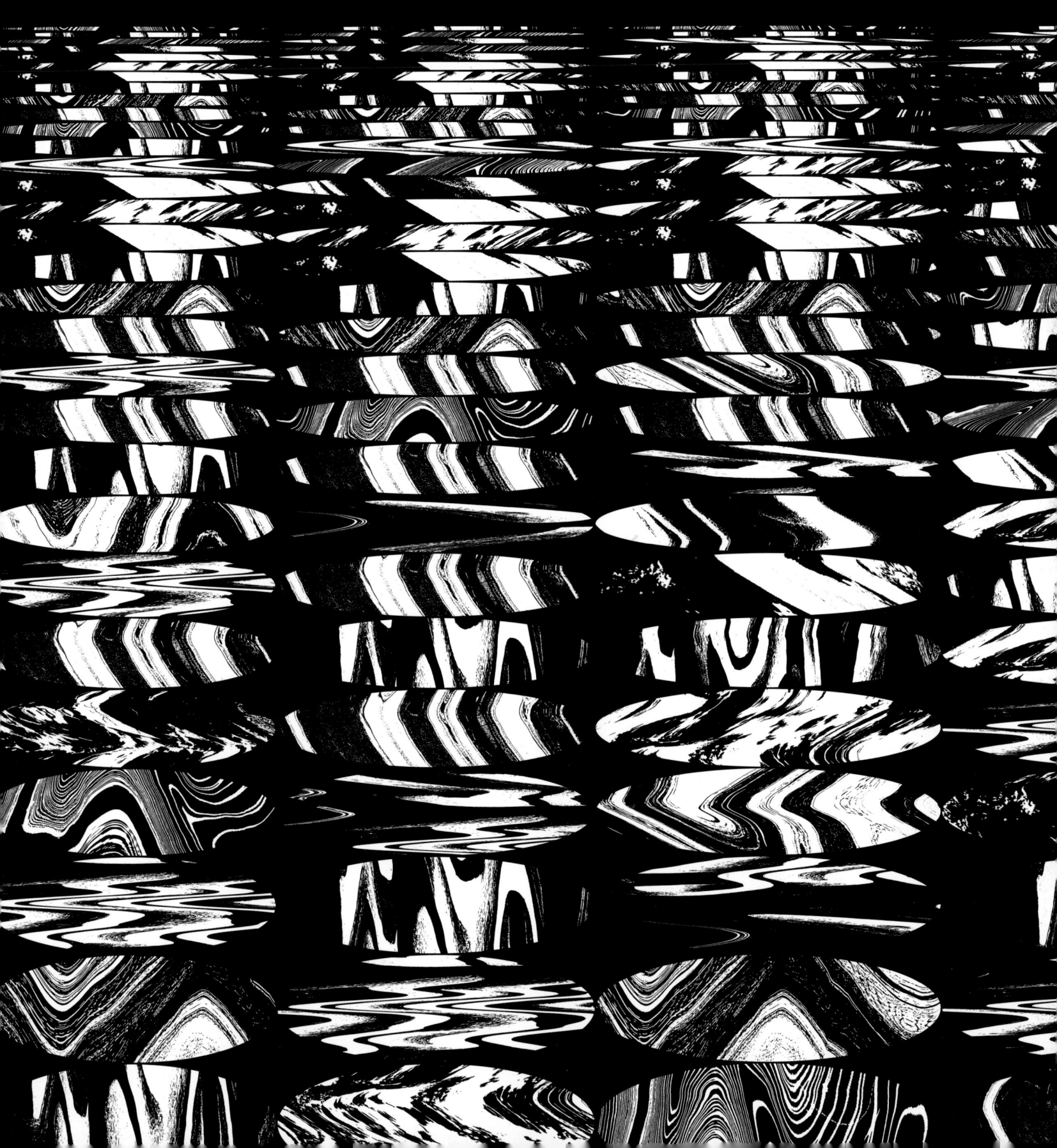

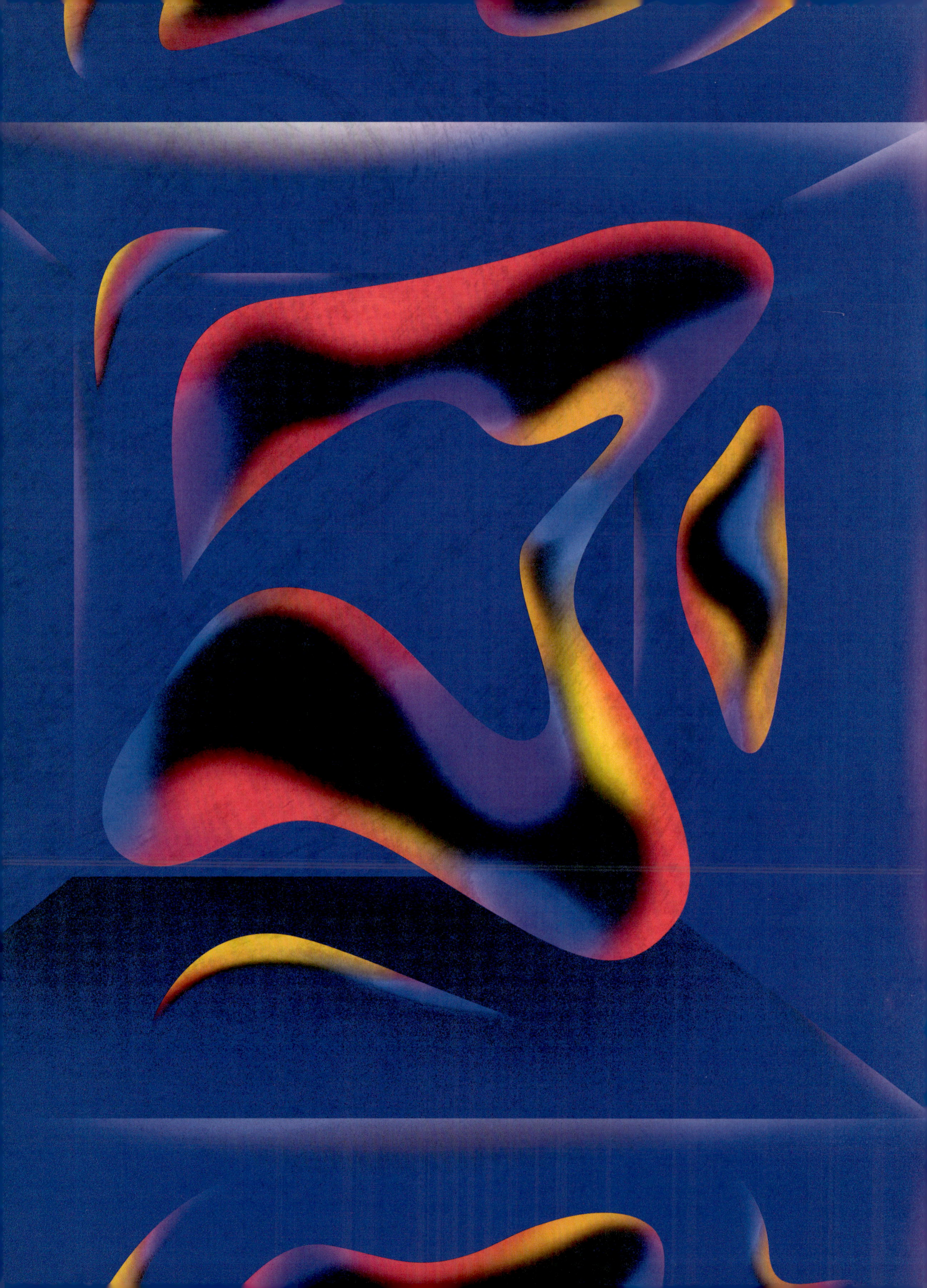

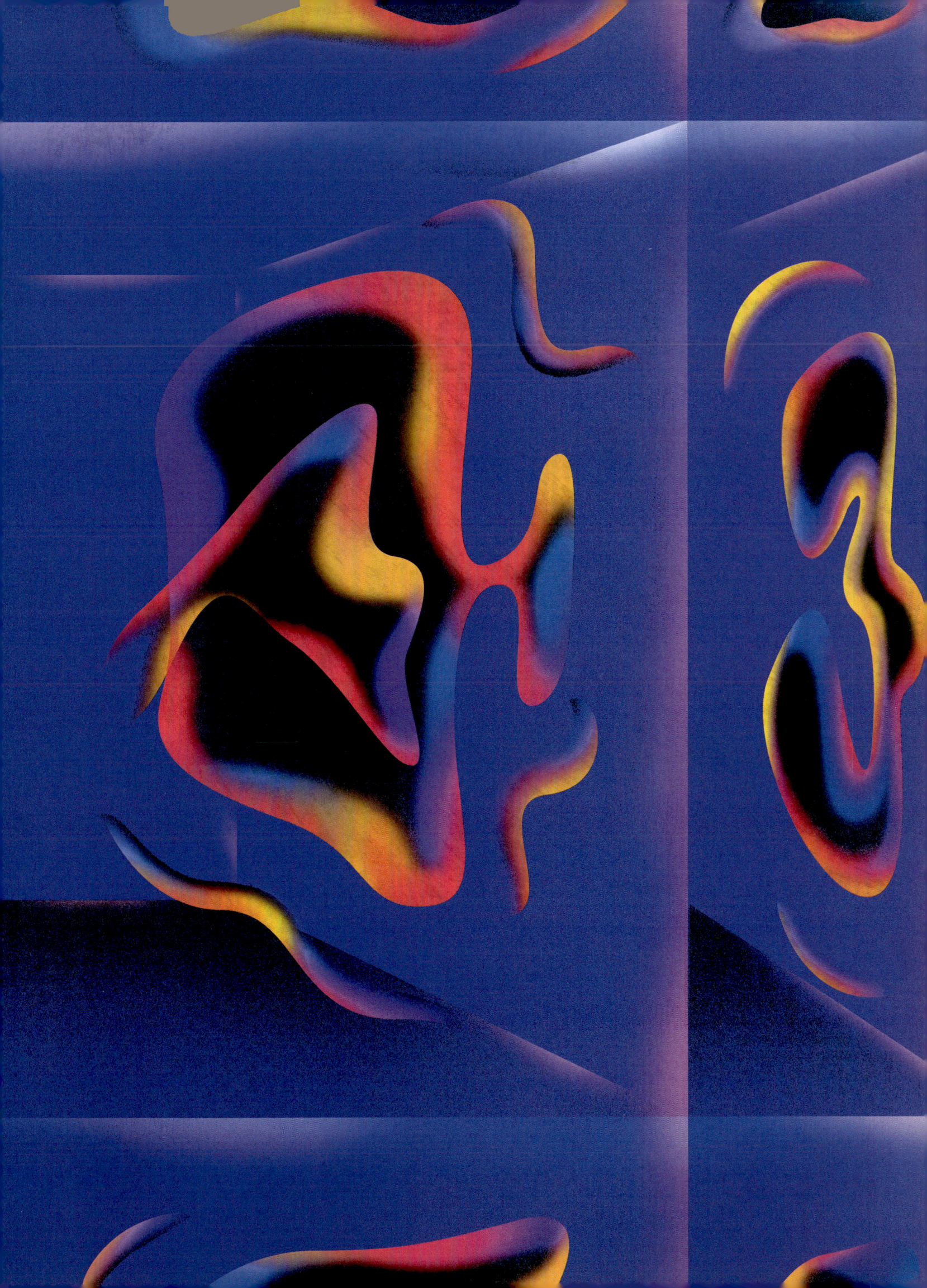

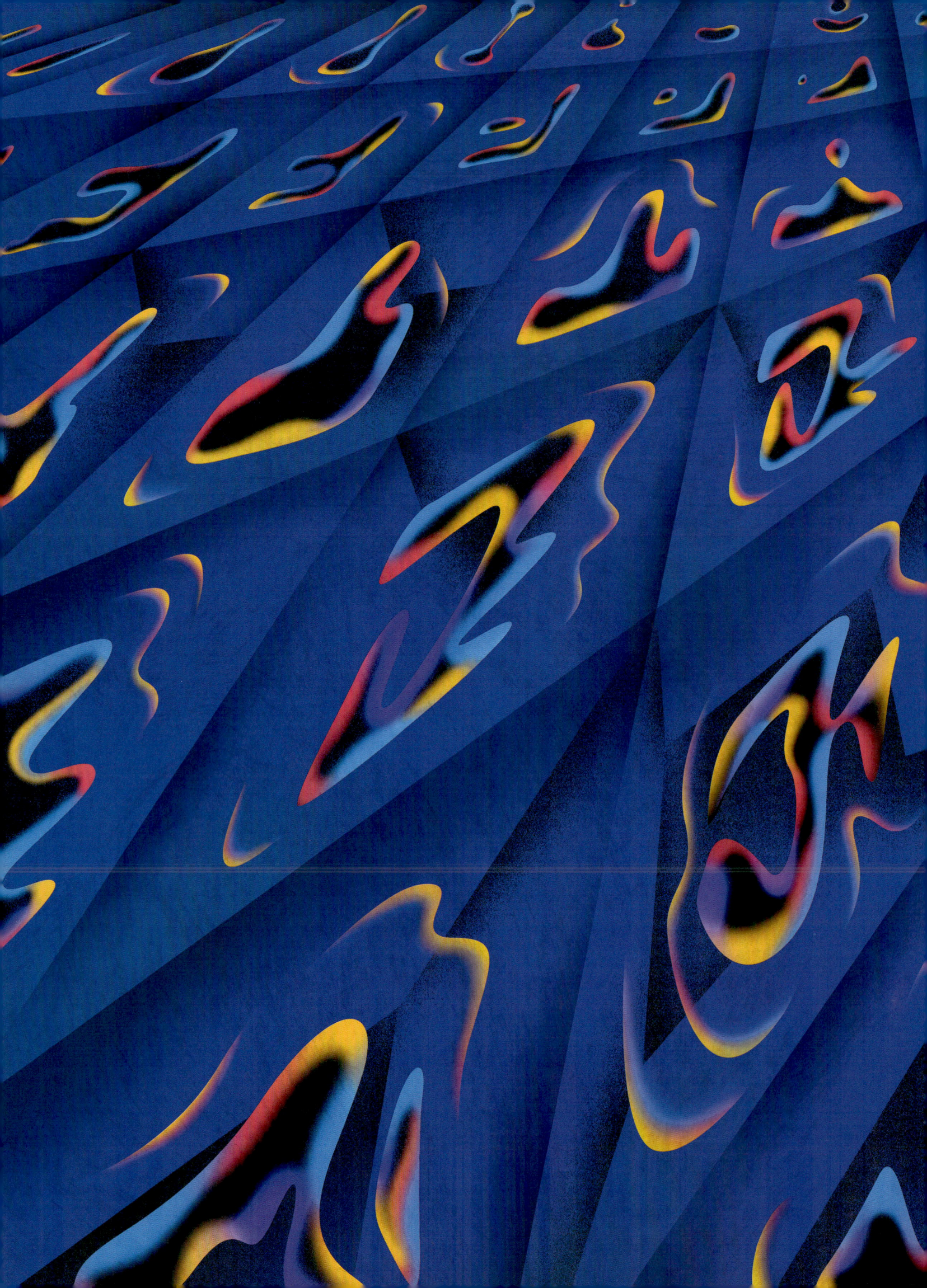

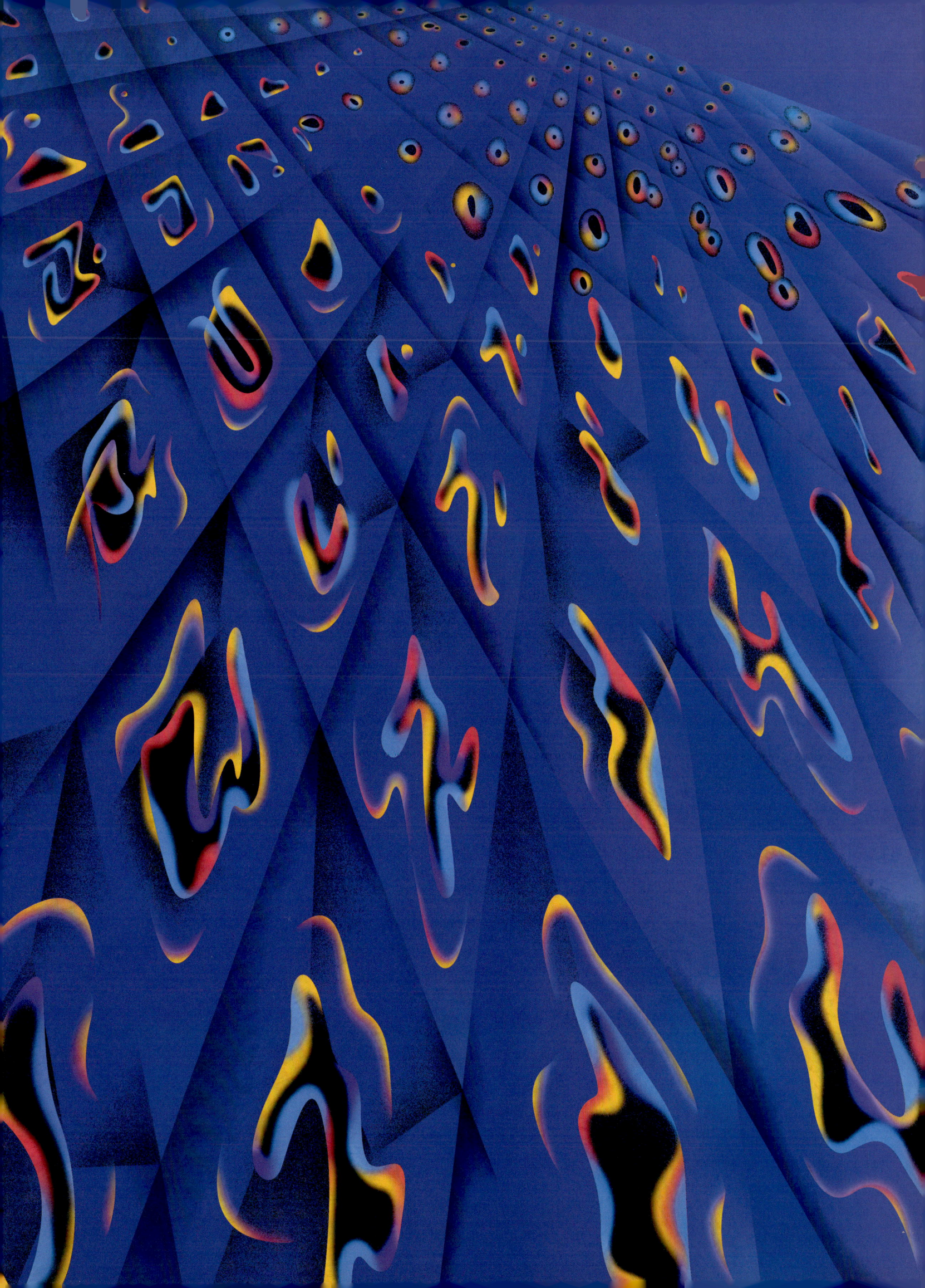

Bubble

In 1980, the cosmologist Alan Guth, contemplating the impeccable smoothness and flatness of the cosmos, realized that space must have blown up and stretched out in an instant, like a balloon. Guth dubbed this explosive growth spurt, the putative opening salvo of the Big Bang, "cosmic inflation." If the theory is right (widely seen as a good bet), then the universe grew from a quantum speck to the size of a beach ball, expanding to something like one million trillion trillion trillion trillion trillion trillion times its initial volume in less than a billionth of a trillionth of a trillionth of a second. It then stopped inflating, expanding gently thereafter, allowing particles to congregate in atoms and larger structures.

Cosmologists who further developed the cosmic inflation theory in the early '80s came to a startling conclusion: once it starts, inflation doesn't stop—not everywhere, at least. Globally, space blasts apart from itself at an exponential rate forever. But within the eternally inflating fabric, little bubbles form in which space stops inflating. As the theory goes, we find ourselves in such a bubble. There are other bubbles out there in the endless sea, other holes in the inflating Swiss cheese, unreachable across the widening gulf between.

Why do the stable bubbles form? In many models, it's because the vacuum of space, which inflates because it's infused with energy, suddenly drops to a lower-energy state, like a ball resting on a table suddenly getting knocked to the floor.

And what are other bubble universes like? Unlike the quilted multiverse, which differs from patch to patch only in how matter is arranged, bubbles in the inflationary multiverse potentially have different elements and governing laws of physics, so the infinite possibilities get even wilder. In many bubbles, space, while it doesn't exponentially inflate, does still expand too fast for stars and galaxies to coalesce, rendering these bubbles lifeless. Other bubbles, like ours, thanks to lucky values for certain physical constants, wound up starry, balmy, livable.

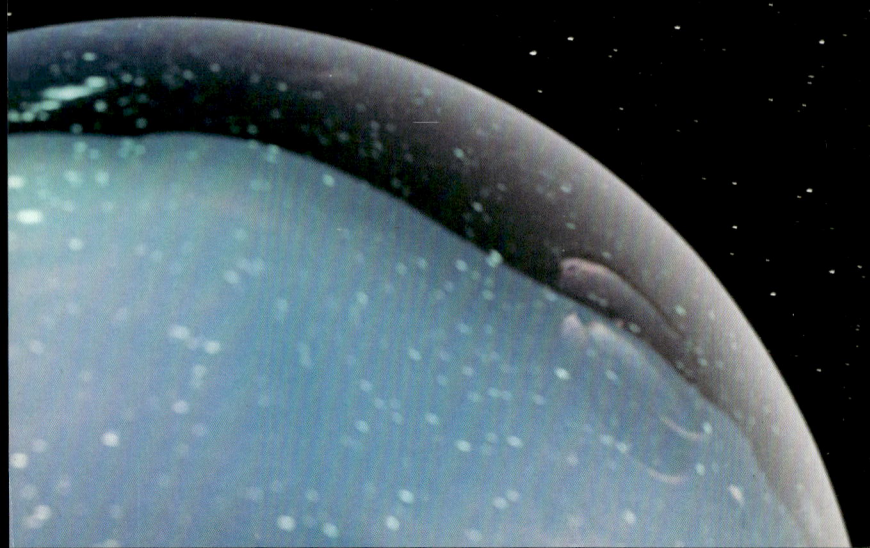

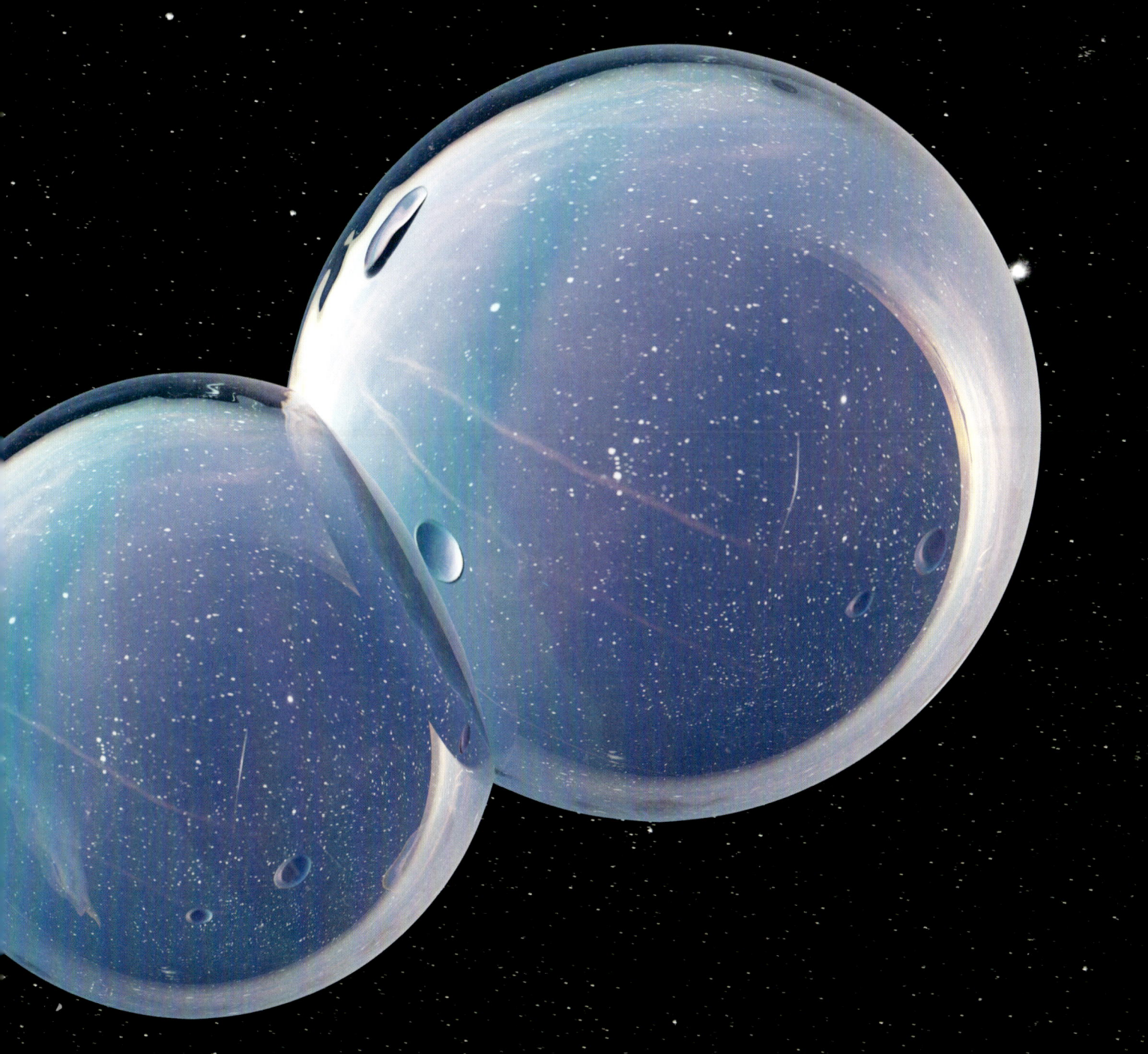

This bubble scenario is accused of being untestable metaphysics. But cosmologists do attempt to study whether the inflationary multiverse explains the properties of our universe. They try to determine whether the particular set of constants that shape our bubble should be typical among the subset of imagined bubbles that could harbor life. If our bubble has features that are statistically common among habitable (and observable) universes, this would serve as a kind of evidence for the inflationary multiverse.

The problem is that when there's an infinite amount of something—in this case, bubble universes—it's unclear how to gauge the proportion of what will be one way versus another. This prevents cosmologists from assessing whether the properties of our universe might be common or rare. Guth put the problem this way: "In a single universe, cows born with two heads are rarer than cows born with one head." But in the multiverse, "there are an infinite number of one-headed cows and an infinite number of two-headed cows. What happens to the ratio?"

Every once in a while I think: By God, we are studying things that we can never physically touch.

We sit on this miserable little planet in a midsized galaxy, and we can characterize most of the universe. It is astonishing to me, the immensity of the situation, and how to relate to it in terms we can understand. — Professor Garth Illingworth

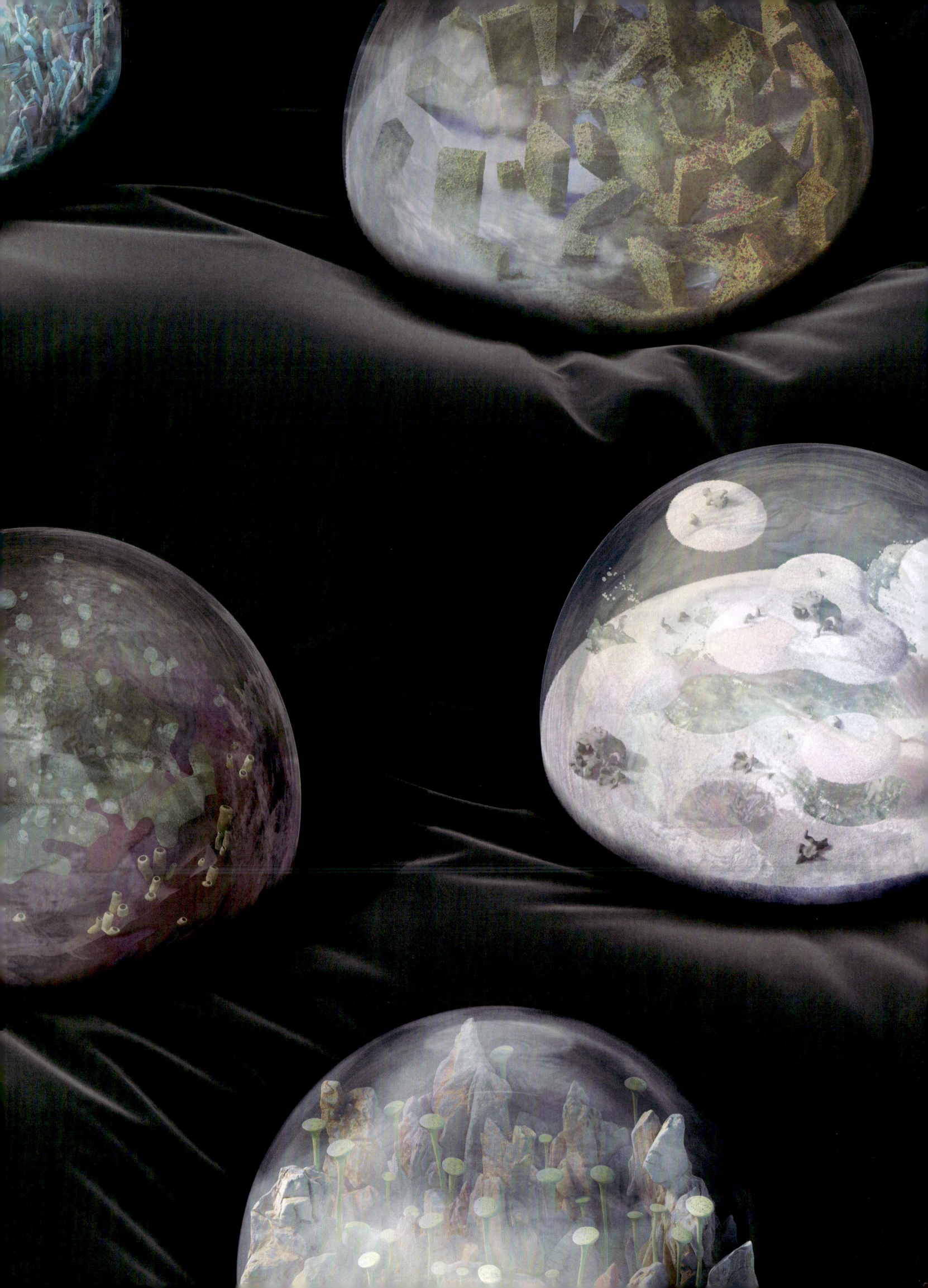

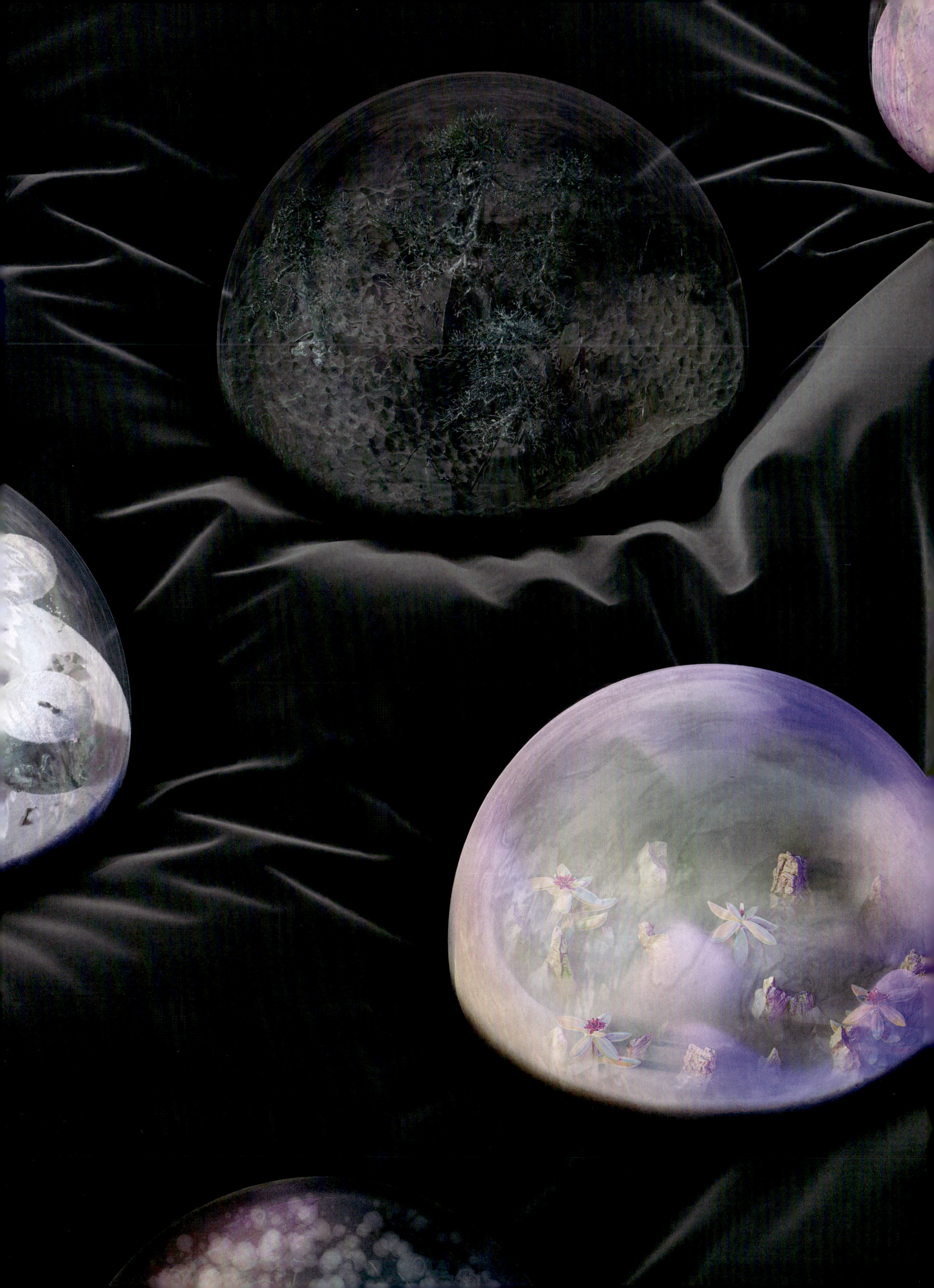

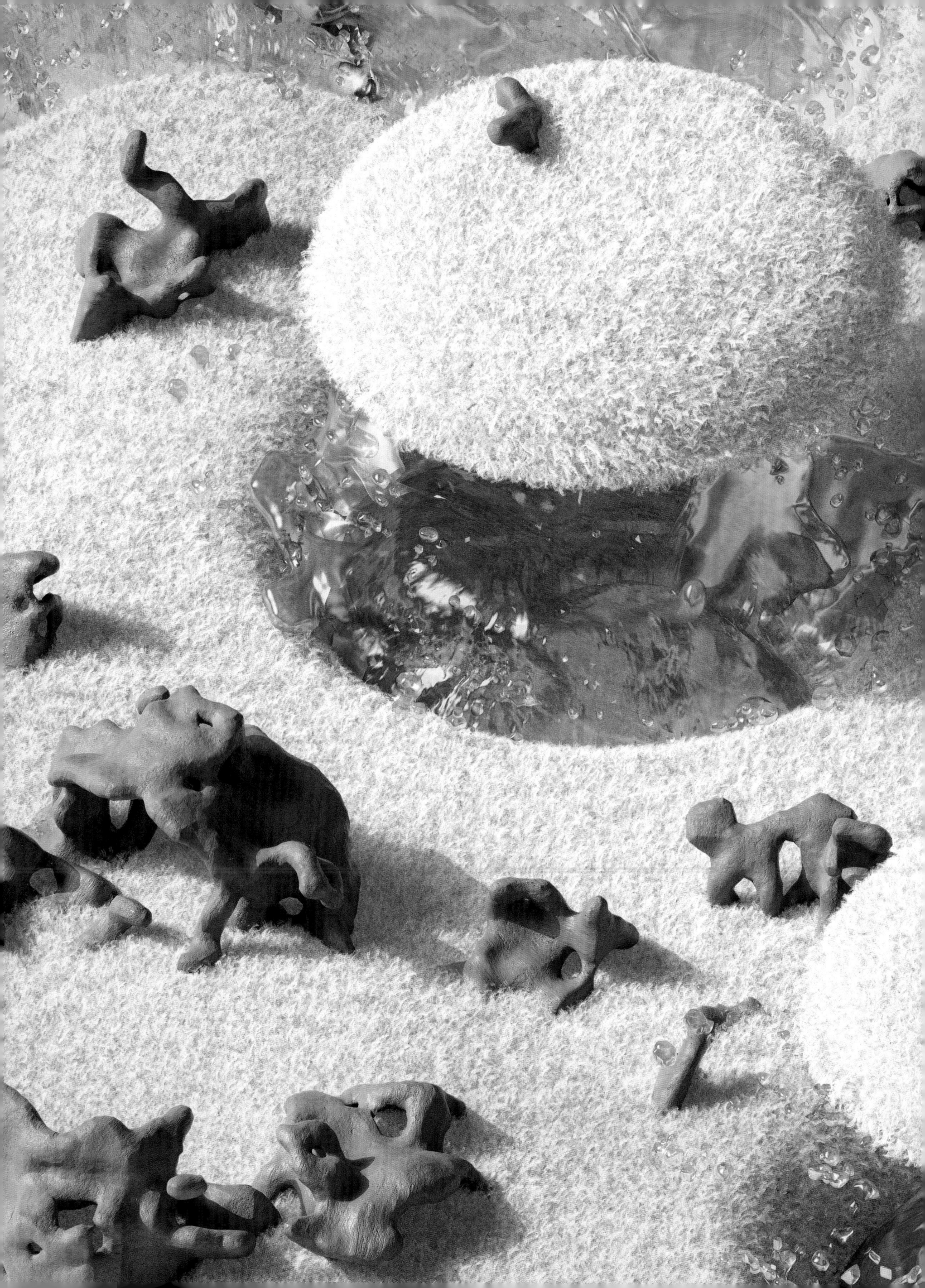

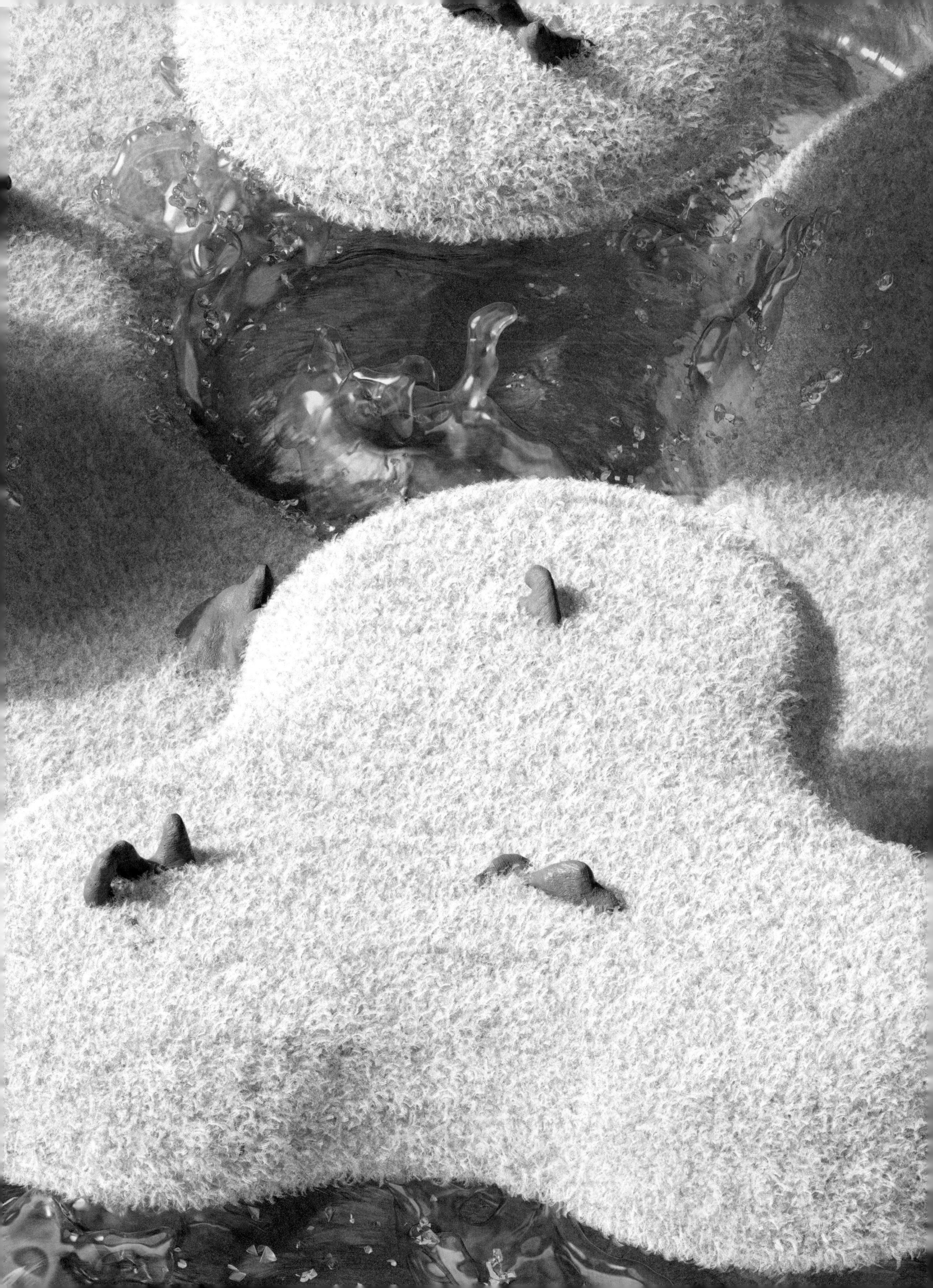

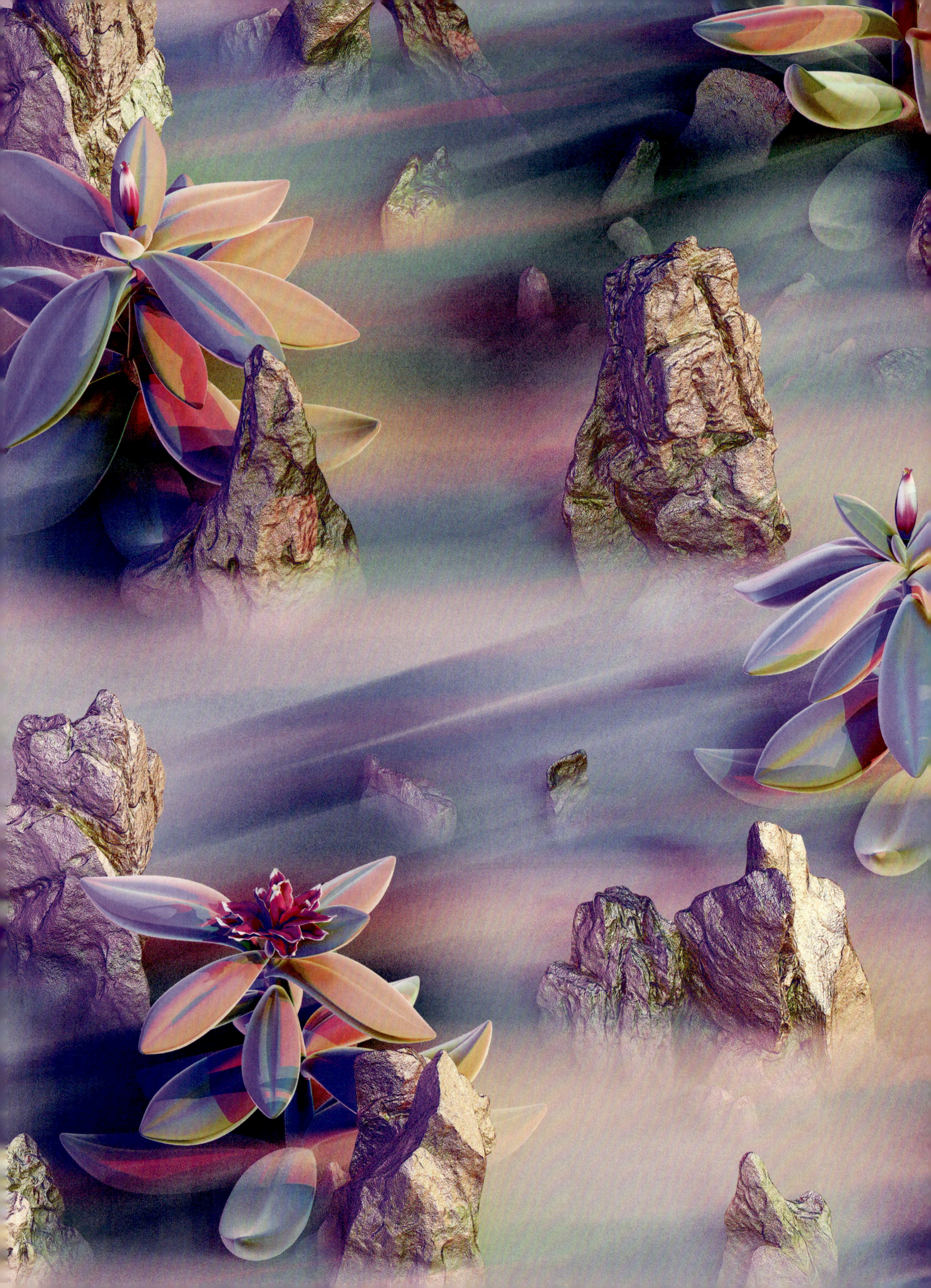

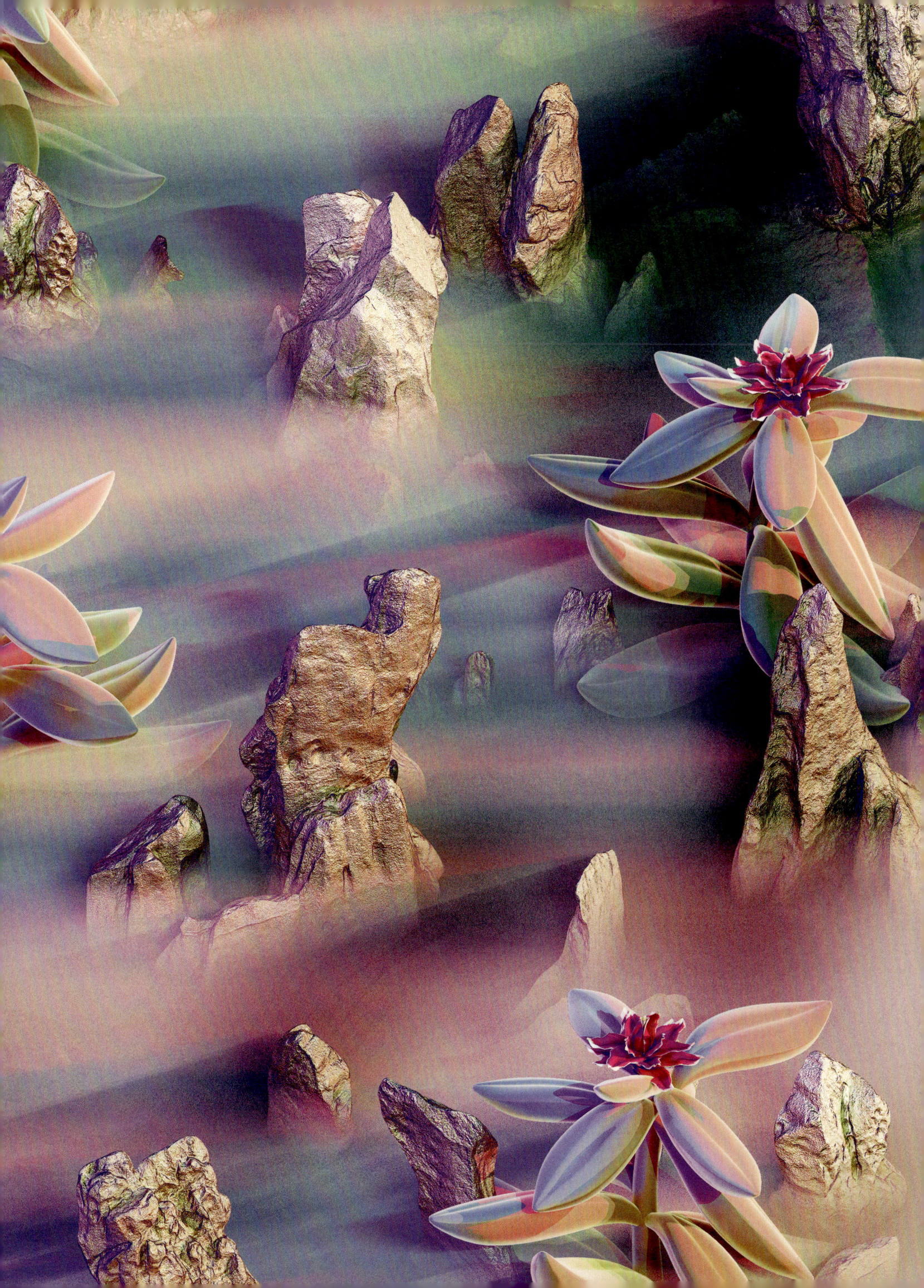

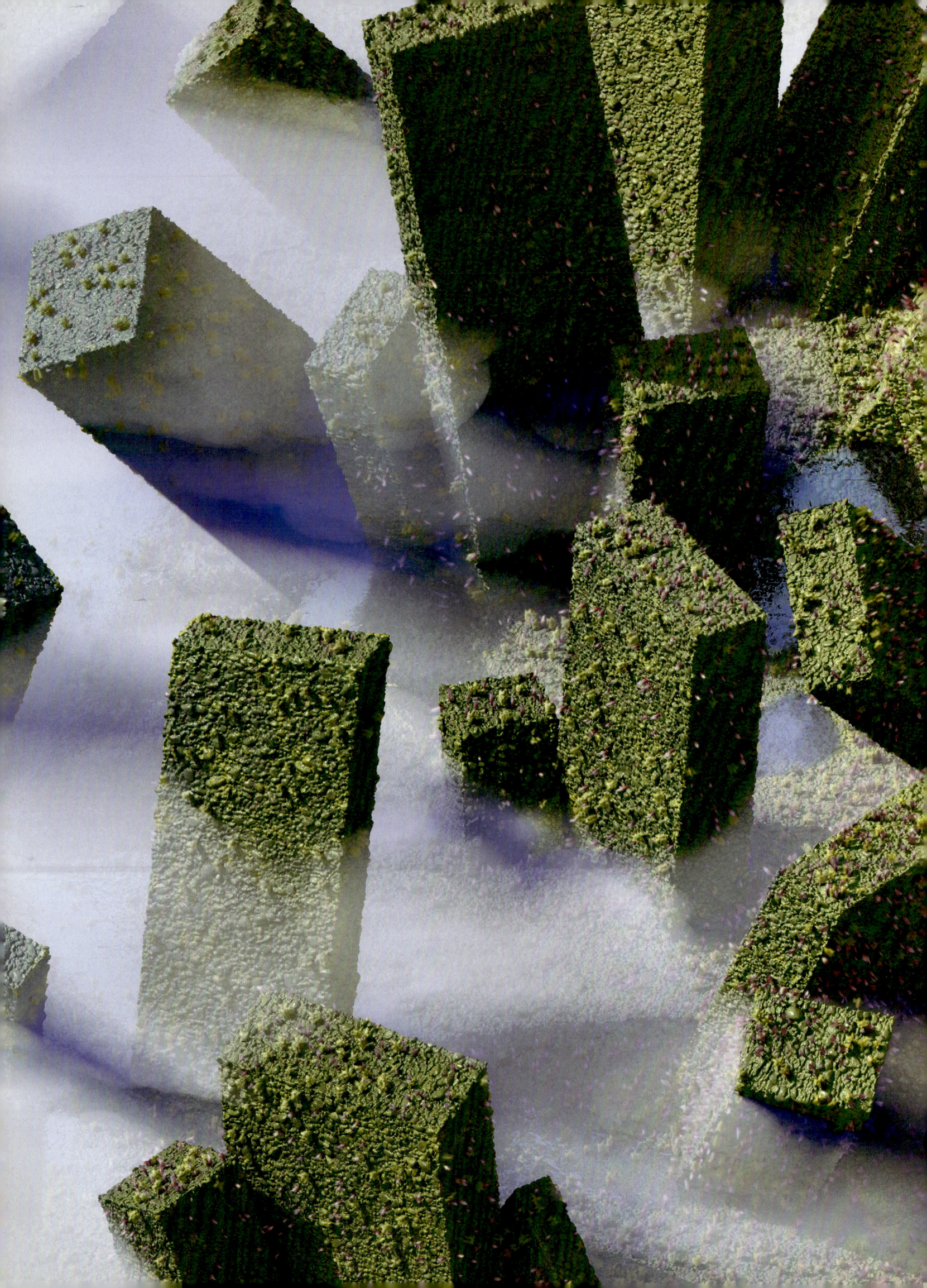

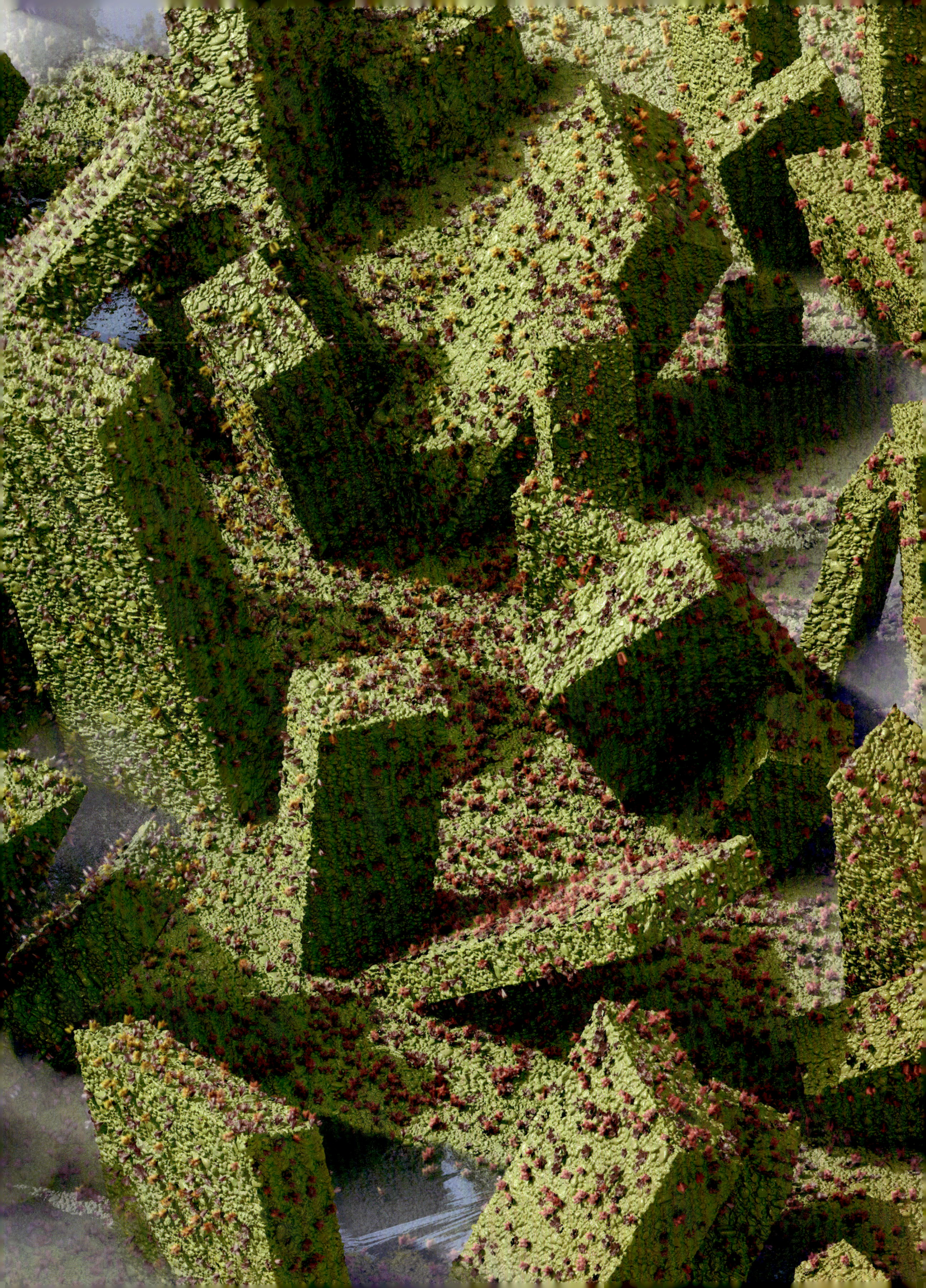

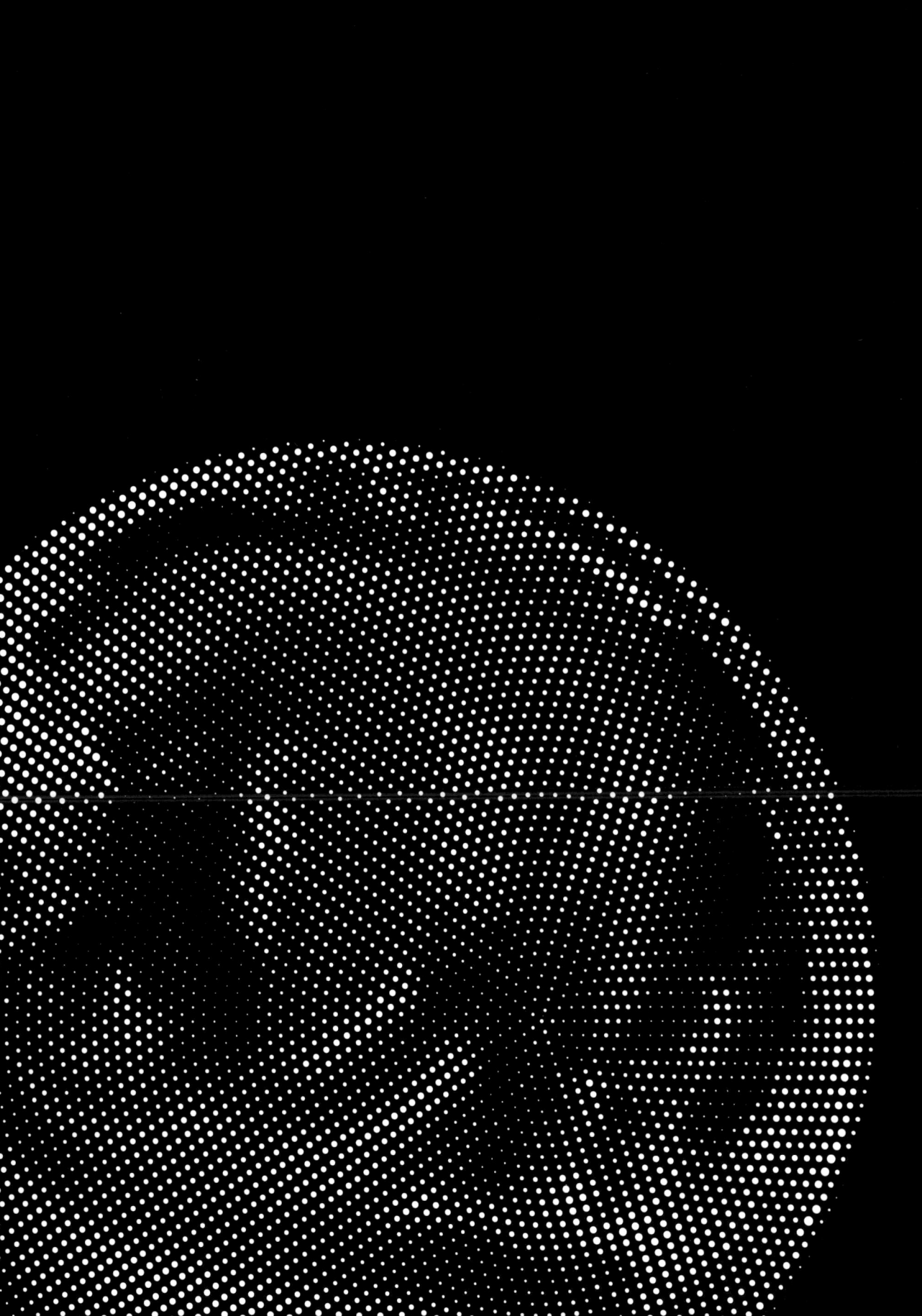

Tiny Gods
by Monique Mitchell

(I am) in 6th grade
Mrs. Smith writes (in chalk,
ignoring the perfectly fine dry erase marker)
"Today's lesson: The Big Bang"
Samantha Suthers storms out the room announcing
"Science is smaller than my God"
Maya, to my left, blows her baby pink gum into a universe
Until—bang!
Her face is a bubblicious delight
She is kicked out, for defiance.

(I am) a sophomore at uni
Professor Gardella writes (in chalk)
"Maya"
The Great Illusion.
Magic trick of the Gods.

He says, the Hindus believe we are all dreaming
None of what we sense is real
Not even this poem (there isn't even a rhyme, is what the existentialists say)
Not even the I, (I am)
Can eye be certain of

Monks spend their lives
Intent on waking up from the dream
Self-help authors insist on creating one's own

This non-poem goes nowhere
Like a wanderer on Samsara's wheel
Clasping at nirvana
Except to say:
We are 8 billion fractals blowing life into a larger One
Expanding and contracting like a sixth grader's stick of bubbaliscious bubble gum
Until bang!
the tiny Gods forget who they are,
and dream again

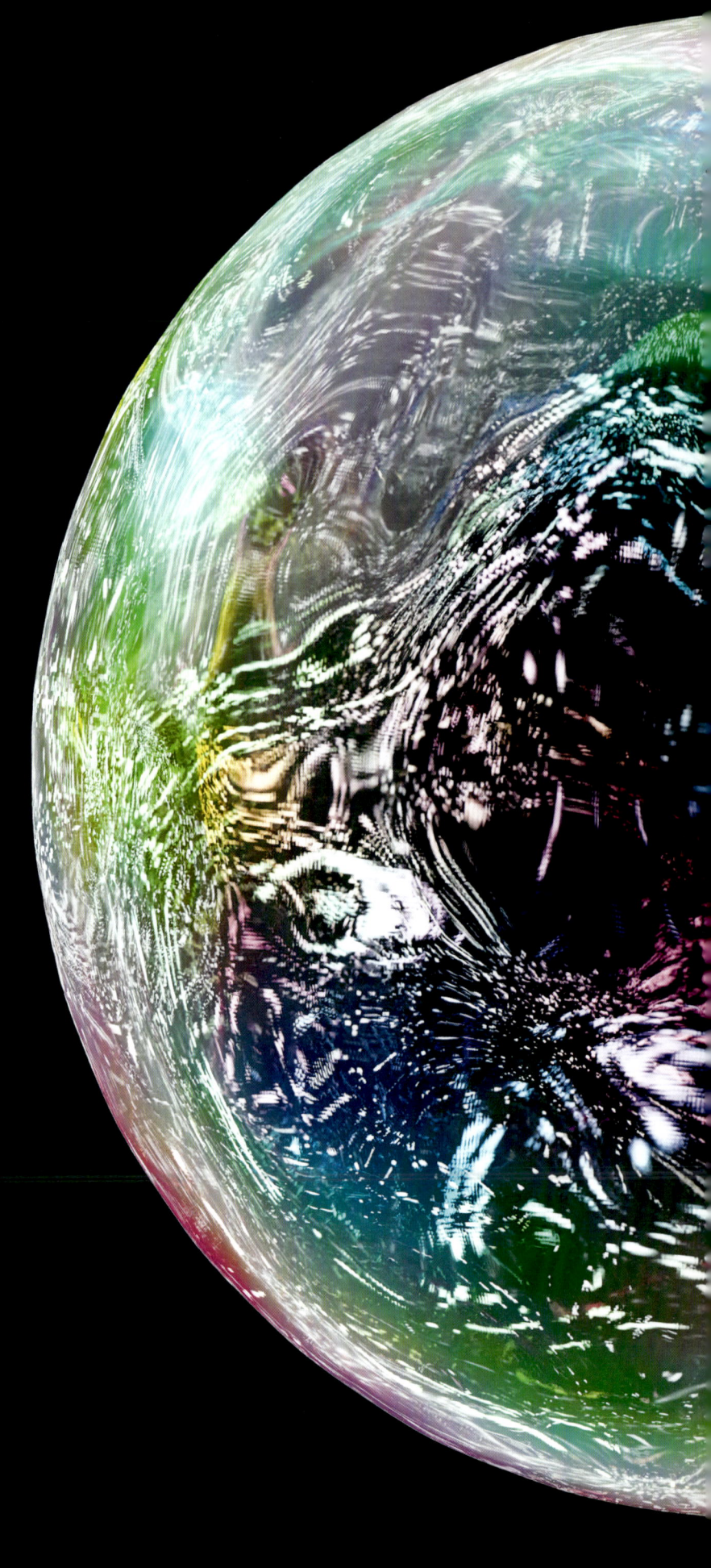

Bubble

POP

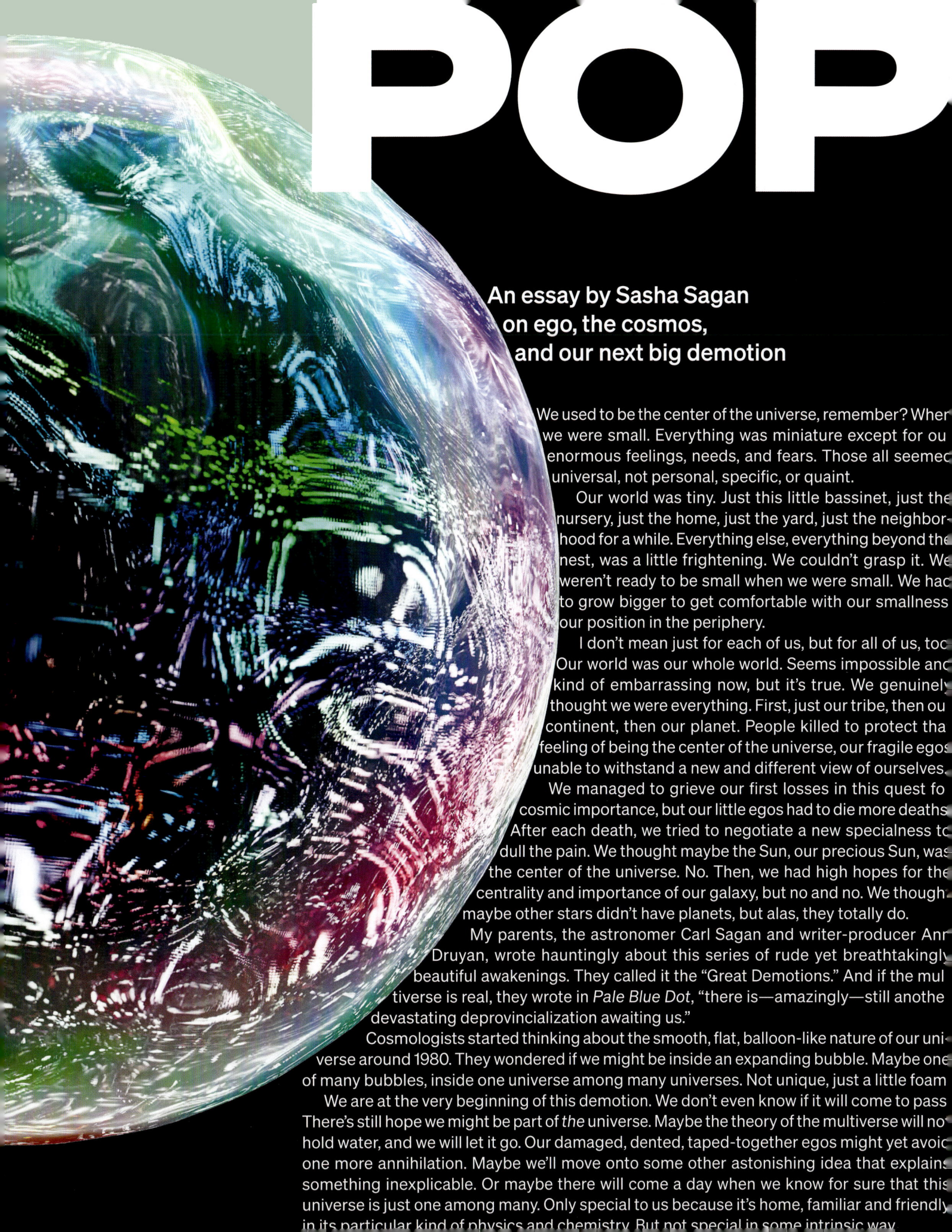

An essay by Sasha Sagan on ego, the cosmos, and our next big demotion

We used to be the center of the universe, remember? When we were small. Everything was miniature except for our enormous feelings, needs, and fears. Those all seemed universal, not personal, specific, or quaint.

Our world was tiny. Just this little bassinet, just the nursery, just the home, just the yard, just the neighborhood for a while. Everything else, everything beyond the nest, was a little frightening. We couldn't grasp it. We weren't ready to be small when we were small. We had to grow bigger to get comfortable with our smallness, our position in the periphery.

I don't mean just for each of us, but for all of us, too. Our world was our whole world. Seems impossible and kind of embarrassing now, but it's true. We genuinely thought we were everything. First, just our tribe, then our continent, then our planet. People killed to protect that feeling of being the center of the universe, our fragile egos unable to withstand a new and different view of ourselves. We managed to grieve our first losses in this quest for cosmic importance, but our little egos had to die more deaths. After each death, we tried to negotiate a new specialness to dull the pain. We thought maybe the Sun, our precious Sun, was the center of the universe. No. Then, we had high hopes for the centrality and importance of our galaxy, but no and no. We thought maybe other stars didn't have planets, but alas, they totally do.

My parents, the astronomer Carl Sagan and writer-producer Ann Druyan, wrote hauntingly about this series of rude yet breathtakingly beautiful awakenings. They called it the "Great Demotions." And if the multiverse is real, they wrote in *Pale Blue Dot*, "there is—amazingly—still another devastating deprovincialization awaiting us."

Cosmologists started thinking about the smooth, flat, balloon-like nature of our universe around 1980. They wondered if we might be inside an expanding bubble. Maybe one of many bubbles, inside one universe among many universes. Not unique, just a little foam.

We are at the very beginning of this demotion. We don't even know if it will come to pass. There's still hope we might be part of *the* universe. Maybe the theory of the multiverse will not hold water, and we will let it go. Our damaged, dented, taped-together egos might yet avoid one more annihilation. Maybe we'll move onto some other astonishing idea that explains something inexplicable. Or maybe there will come a day when we know for sure that this universe is just one among many. Only special to us because it's home, familiar and friendly in its particular kind of physics and chemistry. But not special in some intrinsic way.

And what if universes start and end all the time? Each inflating at different rates from nothing into seemingly everything, some popping before so much as a single star can form inside them. Others making a home for galaxies and supernovae and life and things we cannot yet imagine. What if there are different physical laws in different universes? Does this mean that the immutable laws, like the speed of light, the formula that determines the mass of an object, Newton's law of universal gravitation, the ones we believed were written in stone—unbreakable by anyone or anything anywhere—are just local ordinances?

How do we keep from popping? Stable bubbles last the longest. We are slow in our expansion, growing just a little bit at a time, protecting ourselves from the sudden shock of vastness. Does that make us rare? Or are there trillions of universes approximately like ours?

"Infinity sabotages statistical analysis," writes science journalist Natalie Wolchover. "In an eternally inflating multiverse, where any bubble that can form does so infinitely many times, how do you measure 'typical'?"

When I think of this, of how it sometimes feels to imagine it, I think of someone with whom I once briefly worked. She told me a story about going on a field trip to the planetarium in high school. The visceral, vivid awareness that we are floating in space, emptiness in all directions, was simply too much for her. She had to get up and go sit in the school bus alone, because the scope of the universe made her too uncomfortable. Our smallness was intolerable. Our insignificance was unbearable. Rather than let the existential crisis wash over her, rather than feel those feelings, she left. I picture her there, on the cusp of adulthood, suddenly unmoored. I know that tightening in the chest, the panic. She must have thought by leaving the planetarium, she could avoid thinking about it all. In the near term, perhaps it did the trick. But it did not unring the bell, because decades later, she was still thinking about it. My very having heard the story is evidence of that.

By the time this story was told to me, the teller and planetarium-escaper was middle-aged. She did not tell it to me in the context of how much she'd changed since. It was instead delivered as a point of principle: she did not dive into the deep end and that was that.

While I myself tend toward the deep end, I also have had these flashes of sheer terror at the magnitude of it all. I suspect we all have. It's not just teenagers at the planetarium who feel overwhelmed at our tininess. According to Wolchover, at least one physicist has reported feeling physically ill upon first hearing about "eternal inflation," one of the possible key concepts at the heart of a bubble-style multiverse.

The idea of a multiverse, bubble or otherwise, is to some scientists, outrageously controversial: untestable metaphysics, an attack on everything sacred from Einstein to a general sense of order.

How do we know when a new idea is controversial, because it's just different, upsetting, and a little scary—rather than because it's just plain wrong? We are not always so good at separating our feelings from our facts.

There is resistance. Many physicists insist the nature of our universe must be, well, universal, that any other system is unthinkable and impossible. And that may be so. But the multiverse camp is growing.

If that growing camp is right, maybe out there in the multiverse there is a universe where everything is exactly the same as it is in this one. The same galaxies, the same solar systems, planets, and suns. The same solitary, lovely cratered Moon revolving around the same Earth. The same conditions for life here. The same lucky breaks. The same one-celled organisms. The same dinosaurs. The same asteroid. The same series of events leading to snails and octopuses and amoebas and porpoises and chipmunks and *Homo habilis*. Then the same *Homo erectus*, Neanderthals, *Homo floresiensis*, the same *Homo sapiens*. All of us, the exact same. Except for one: the young woman at the planetarium. In this universe, she faces her fears. She stays. She feels our smallness, and after a little while, her perspective profoundly changes. Her fears about her final grades, some drama with a friend, some party she wasn't allowed to attend, her future, the meaning of life and death—all lift. She is suddenly swept up in a sense of sublime joy that she is part of something grander and more spectacular than she ever imagined.

She lives a freer, braver, fuller life. And maybe when she is wise and wizened she reads there is new evidence to suggest our universe is but one foamy bubble among countless others, and she laughs with delight at the endlessness of possibility and pure ecstasy and unlikelihood of existing anywhere at all.

If we are someday able to discern whether the bubble theory of the multiverse is, in fact, the nature of the next tier of our reality, how will we feel? Will we long to cross into the next bubble the way we yearn to know Mars? Could we ever get there? Can we break the bounds of our universe? Is it too dreadful to even think about? Are we happier here, in our one universe? In our empty school bus? Alone?

We can try to put it out of our minds, but only for so long. The scientific method will, in time, lead us somewhere beyond our needs and fears. Maybe to infinite bubble universes, maybe somewhere else. But whatever it is, we will, I hope, in time, come to grips with another demotion. Eventually arriving at a place where the pain is gone and we are at comfortable peace with our true place

No matter how far you go, you can go further.

Andrei Linde

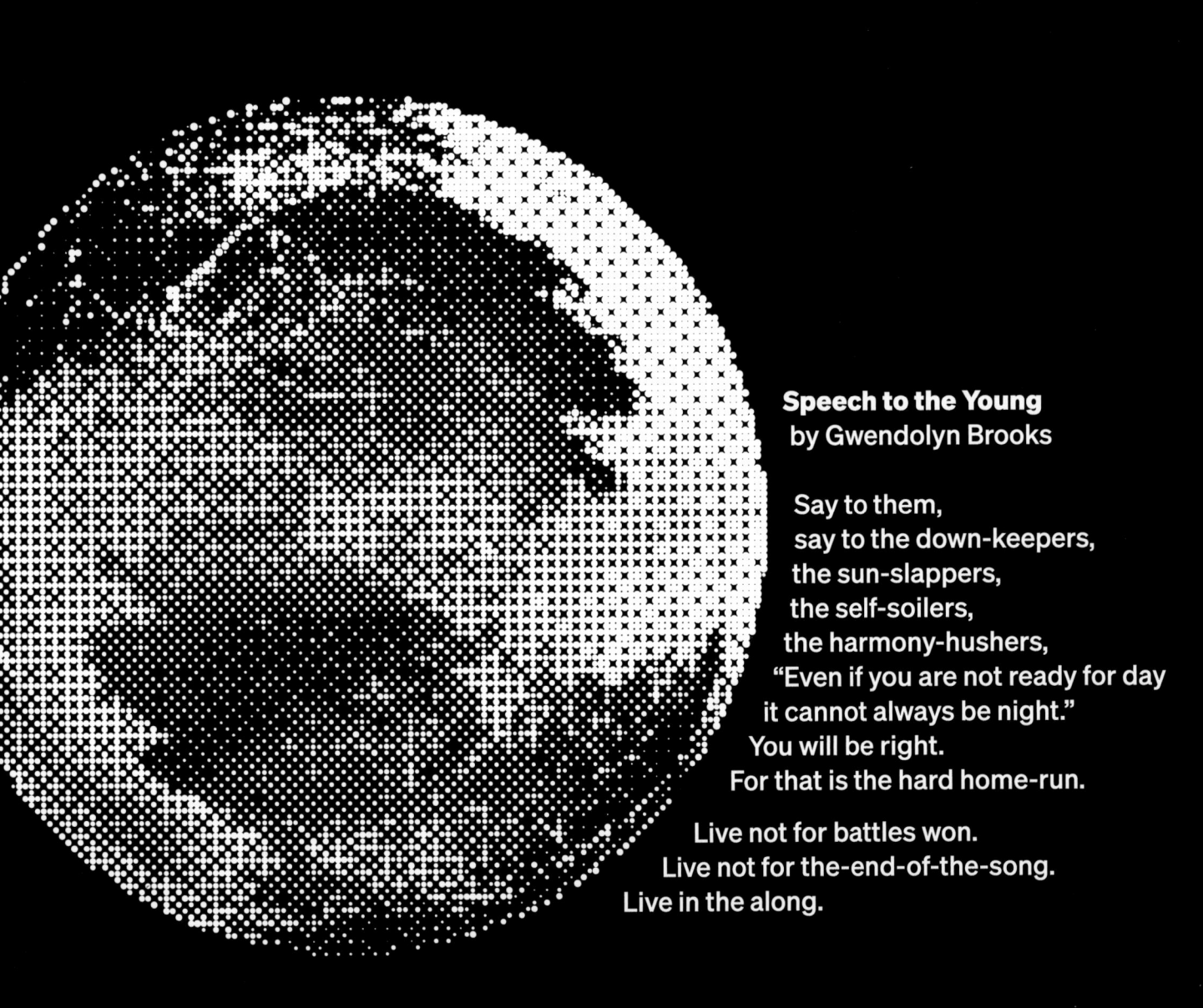

Speech to the Young
by Gwendolyn Brooks

Say to them,
say to the down-keepers,
the sun-slappers,
the self-soilers,
the harmony-hushers,
"Even if you are not ready for day
it cannot always be night."
You will be right.
For that is the hard home-run.

Live not for battles won.
Live not for the-end-of-the-song.
Live in the along.

Chapter 4

Surface

What if everything we experience is just the surface of all there is? That's the gist of the braneworld scenario, an outgrowth of string theory.

To imagine our 3D reality as a surface in some higher-dimensional space, first imagine what it would be like to live on a 2D surface within our 3D universe. You could be a microbe that lives on skin, for instance. The microbe can move forward or backward or side to side, but it can't go (and lacks even an awareness of going) up or down. Confined to a sheet, it's ignorant of the higher-dimensional space in which it's embedded.

String theory famously posits that elementary particles are made of minuscule vibrating strings. Along with these 1D strings, string theory also postulates 2D surfaces known as membranes, or "branes." There are also 3D and higher-dimensional branes. Researchers have theorized that the entire universe is a 3D brane, one that, like skin, is embedded in higher-dimensional space. If we inhabit a brane, we can have no conception of the orthogonal dimensions that make up the rest of reality nor, in all likelihood, of other braneworlds that may reside there. Other branes could be separated from ours by a microscopic distance, yet whatever dramas have unfolded there we'll never know.

Unlike the three familiar spatial dimensions, any extra ones must be extremely small, because we don't see them. You can think of an extra dimension as a tiny loop that exists at every point in our 3D space. If you could rotate around the loop, you'd leave our universe and explore additional places in the landscape. In practice, though, we can't rotate out of our brane. We're stuck on this surface, like skin bacteria.

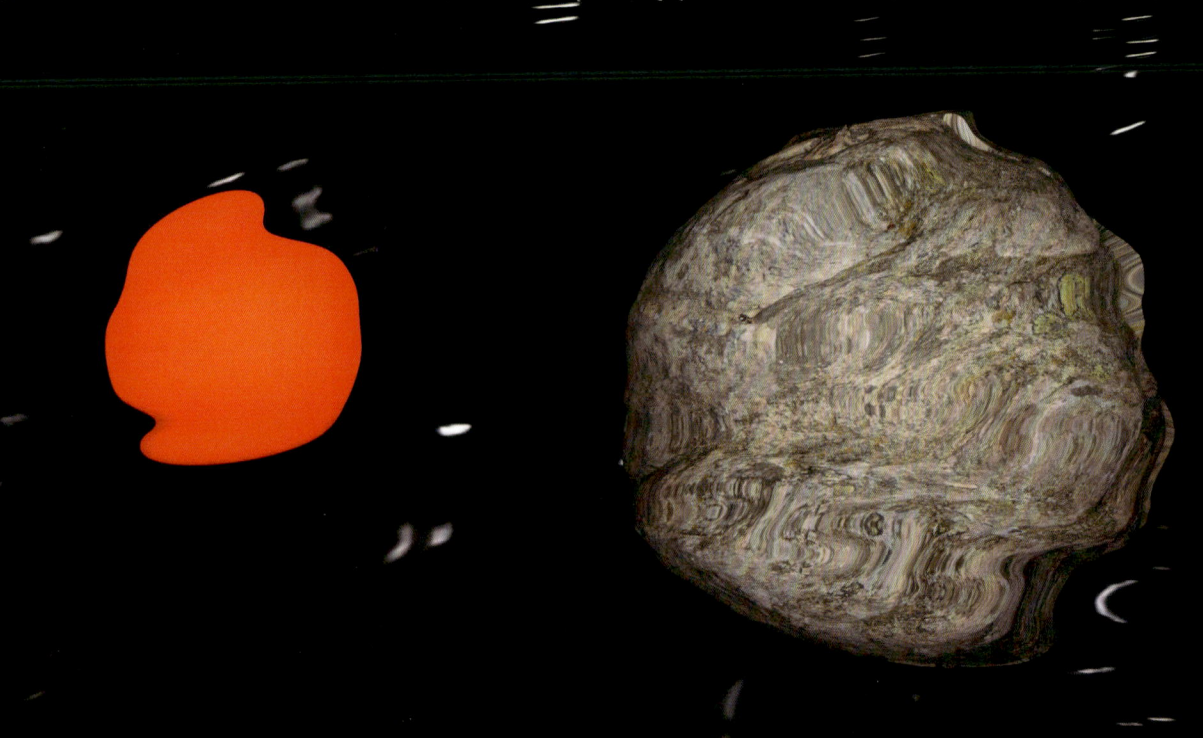

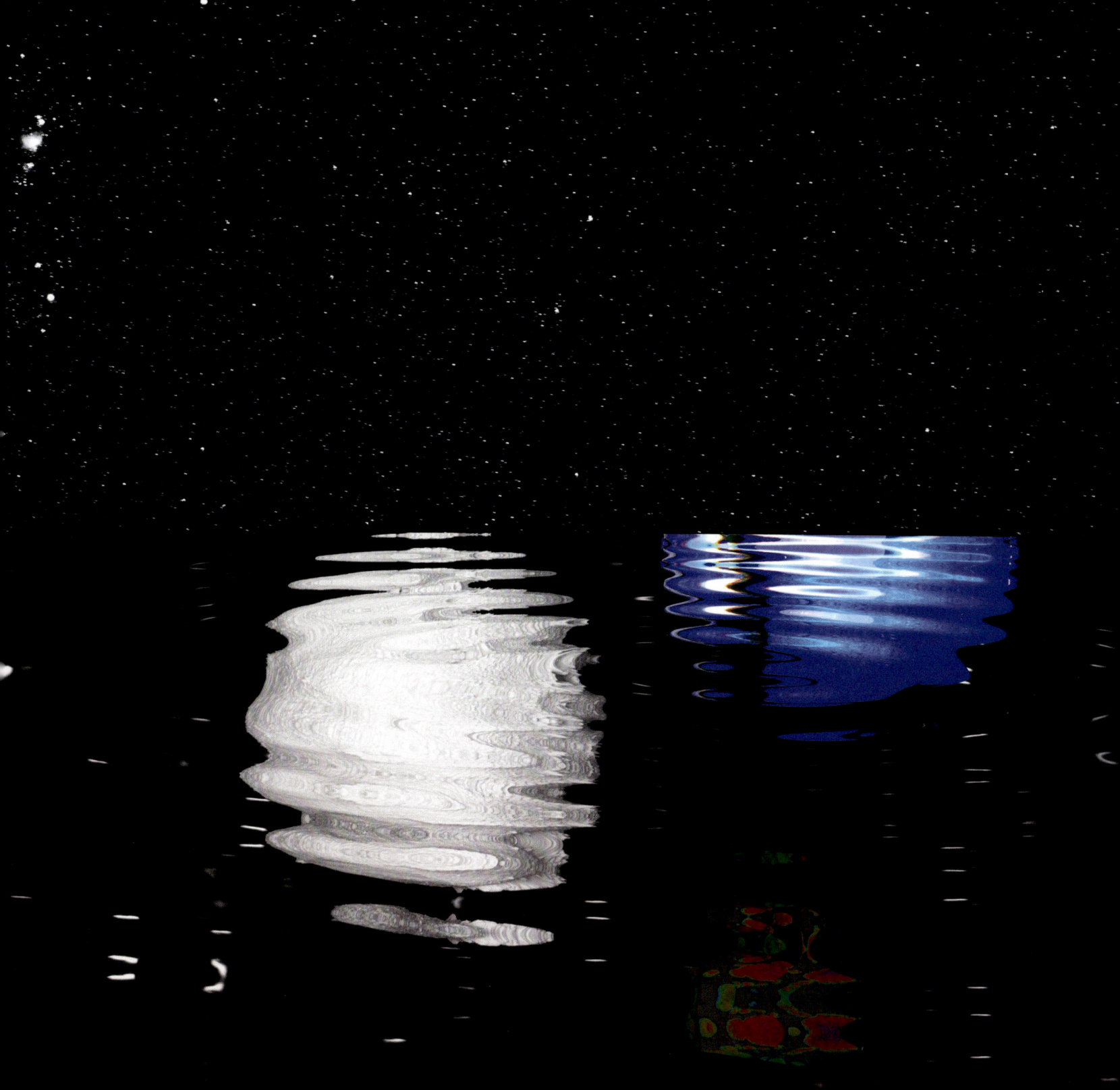

Stop acting so small.

You are the universe in ecstatic motion.

Rumi

I think that's when I realized, says Richard, that the things I can endure are only just the surface of what I can't possibly endure.

Like the surface of the sea? asks Khalil

Actually, yes, exactly like the surface of the sea.

Jenny Erpenbeck, *Go Went Gone*

Daniel and Daniel with David[1]

An interview between two filmmakers and their favorite neuroscientist, David Eagleman

By Daniel Scheinert and Daniel Kwan

1. We added these annotations to help explain some of the stuff we reference from David's books.

We've been avidly reading work by neuroscientist, professor, writer, extrasensory technology inventor, and frequent Radiolab guest David Eagleman for years now. In the summer of 2021, we had the opportunity to let our brains pick his brain and discuss the themes of the multiverse with him.

Though Eagleman's work has rarely, if ever, touched upon the subject of the multiverse, we knew that, much like us, he is interested in the gap between what is known and what is unknowable, and the ways that art and science attempt to bridge that gap. We also thought that a conversation about the multiverse with him would be fascinating and fun. We were right.

—Daniel Kwan and Daniel Scheinert

DANIEL KWAN

One thing we're always thinking about is the stories we tell and what that says about the times that we're living in. Right now, it feels like the multiverse is everywhere in pop culture—Marvel, *Star Trek*, *Rick and Morty*. Why do you think that is happening, and what do you think that says about this cultural moment?

DAVID EAGLEMAN

Literature usually follows science. Once something hits the table in science as a strong hypothesis, authors, sci-fi writers, writers of literary fiction, they all want to use it. The same thing has happened with quantum mechanics. It serves as such a powerful literary mechanism to explore possibilities in parallel and imagine that these all exist at once. It's a really powerful way to imagine and illustrate what *could be*, and this is a big part of what human brains do—we're constantly thinking "what if?"

DANIEL SCHEINERT

As a scientist, do you find that exciting or frustrating? Do you feel like storytellers often misrepresent the science, or do they help bring it to more people?

EAGLEMAN

There's a sense in which the storyteller's job is to get things mostly right, but I don't think it's their job to bring science to the people. It's their job to use whatever mechanism that allows them to tell a story that's new and fresh and important. And by the way, the idea of the multiverse might not be correct—I mean, it might be a good explanation for certain things, but it might not be. Nonetheless, it's a useful way to structure a story. I'm both a writer of literature and a scientist, so I guess I would defend both roles and say neither one is subservient to the other.

SCHEINERT

I think that's what makes your science writing so accessible—it will just become a story about the brain, and then suddenly I understand the science. I feel like you do use stories to get science out into the world.

EAGLEMAN

Story is the only way to get information in there. We're really wired for story.

SCHEINERT

That sort of speaks to Possibilianism,[2] the idea spawned by your book of fiction *Sum*, that an infinite number of afterlives *could* be true, so maybe *all* possibilities are useful thought experiments. Can you talk about what you're proposing with that idea?

Surface

EAGLEMAN
I'm proposing that, in the possibility space, all of the religions are one point to that space. The idea that we die, and there's nothing else, that's another possibility in that space. There are lots of possibilities. My interest has always been in exploring the structure of that space. That's why I wrote *Sum*[3] to shine a flashlight around the possibility space. None of the stories were meant to be taken seriously—they're all funny stories—but it's meant to demonstrate that there's a lot of things that we don't talk about.

KWAN
In one of your talks, you talk about the problem of this false dichotomy: this belief system is right, and this belief system is wrong. Rather, they're all potential data points in this bigger picture that's alluding to something we may not possibly ever understand. It makes me wonder—in your eyes, how do we actually move beyond this false dichotomy and get to that next stage in our higher perspective or collective understanding?

EAGLEMAN
That's a good question. The point of science is to understand. We present a set of hypotheses, then we do experiments on them. With experiments, you can rule out whole parts of the possibility space. You can open up new parts of the possibility space, too, and that's the idea, but you've got to get the whole table set right first.

What generally happens [with belief systems] is that you've got the religions, and you've got atheism, and those are the only two positions people take. It's also what goes on with the political parties, where people entrench themselves in a position that's very limited. The question is, if we're going to apply science [to belief systems], how can we actually really look around and get somewhere? You know, [the answers] might be ten thousand years past our lifetimes, but that's the goal of science: to say, all right, here's what we're exploring. Let's go look for evidence and find things that might point us in the right direction.

> 2. From the official Possibilianism* website: "Possibilianism is a philosophy which rejects both the idiosyncratic claims of traditional theism and the positions of certainty in atheism in favor of a middle, exploratory ground."
>
> * https://www.possibilian.com

> 3. *Sum: Forty Tales from the Afterlives* is a playful collection of short stories in which every story is an efficient, yet mind-expanding thought experiment, each exploring possible versions of the afterlife.

SCHEINERT
I feel like we're inching toward why we thought of you when we thought about the multiverse and why it's in so many stories right now. As a society, we're now staring at the internet and finding just how many different belief systems there are and just how unhelpful a black-and-white story is for understanding the world. Now you can just log on and see a torrential amount of opinions and possibilities out there.

EAGLEMAN
Yeah, that's a cool analogy. This is actually the topic of my next book, which is about how the brain makes an internal model, and we all come to believe that our internal models are correct. Like, "I've clearly got the right view of the world, and everyone who disagrees with me on social media is malicious or ill-intentioned or not bright enough to understand." This is what we all have in our heads. But it's because we're stuck with our own models and we're not particularly good at extrapolating and trying to get out of them.

KWAN
You talk about the importance of the ability for humans to train ourselves to be able to hold opposing or even contradictory ideas in our head at the same time. It's not—at least from my experience and the way I look at the world—an easy thing. It's not the default. As a neuroscientist, do you think there is a way that we can train ourselves as a society to do this? What does that look like?

EAGLEMAN
I generally think this is part of the passage into maturity, and it's somehow not taught. It's certainly not taught to our young people now. Instead, it's that whatever political opinion you have is true and it's right. I think the only way for us to get closer along that path is just making this part of our education as opposed to, "if you feel that way, that must be the truth of the matter."

SCHEINERT
So we're capable of it, but it's maybe not intuitive.

KWAN
And it's not the priority.

SCHEINERT
It was a mind-blowing concept when I found out, embarrassingly late in life, that history is up for debate. That there are people out there arguing about what went down. That was so much more interesting to me than "here's the date that this happened," you know? I wish the concept of opposing viewpoints had been introduced into every discipline earlier. Science is so much more interesting when it's a conversation.

EAGLEMAN
You know what's so interesting about that? Science is always driven by debate and by people saying, "Wait, I don't believe your results. I'm going to do some experiments." That's how everything moves forward—it's two steps forward, one step back, sometimes even as things progress. It's been a very interesting year and a half with the coronavirus [COVID-19], because debate has gotten squelched from the get-go, rather than seeing all the opposing viewpoints and how those work.

KWAN
To play devil's advocate, I feel like part of the reason why we evolved to try to find black-and-white, concrete things is because it's the fastest way to navigate the confusion and chaos, where we have to make decisions. Otherwise, as an organism, you would just flounder and get confused and die. I do think maybe it is about finding that balance. Right now, with the coronavirus, even though it feels wrong for, say, the CDC [Centers for Disease Control and Prevention] to make decisions and say, "This is what is actually happening in the world, and this is what we know based on science," it's just so that decisions can be made. Otherwise, we get stuck in this vortex of everyone talking about the same thing, but coming from completely different points of view.

EAGLEMAN
You are absolutely right. That's exactly the conflict—always. Do you want a clear answer? Or do you want to debate and get to the truth, which takes a longer time?

KWAN
One of the things that you introduced to us early on that really changed the way that we thought about how we tell stories and how we looked at the world was through the concepts of the Umwelt versus the Umgebung...[4]

SCHEINERT
...Umwelt being what we can observe, the observable universe, and Umgebung being what we don't know, which could be immeasurably immense.

KWAN
As someone who thinks about this all the time, we wanted to ask you, what is your wildest conception of the Umgebung? What are some of the crazier things that you have considered might actually be possible?

EAGLEMAN
Well, it's funny because to some degree, I published 40 such ideas in *Sum*. But to give you just one example of something that I'm thinking about now—I'm writing a new book of fiction and it's sort of like *Sum* in its mischievous spirit, but one of the ideas I'm exploring is [that] there are many spatial dimensions, and we live in three of them. So could there be whole civilizations in other spatial dimensions? Like, in dimensions four through six, there are some other civilizations living there. I use this for comedic effect in the book, but you know, who the heck knows? Of course, we wonder if there is life on other planets, but what if there's life all around us in other dimensions that we just can't see? So that's a wacky idea that I've been enjoying for a while.

KWAN
I can't wait to read that book.

SCHEINERT
I loved reading about some of the stranger things you've been exploring with sensing, like your sensory vest[5] for trying to train the body to just understand the stock market by vibrating.

4. On his blog, David writes:

"In 1909, the biologist Jakob von Uexküll introduced the concept of the Umwelt. He wanted a word to express a simple (but often overlooked) observation: different animals in the same ecosystem pick up on different environmental signals. In the blind and deaf world of the tick, the important signals are temperature and the odor of butyric acid. For the black ghost knifefish, it's electrical fields. For the echolocating bat, it's air-compression waves. The small subset of the world that an animal is able to detect is its Umwelt. The bigger reality, whatever that might mean, is called the Umgebung. The interesting part is that each organism presumably assumes its Umwelt to be the entire objective reality "out there." Why would any of us stop to think that there is more beyond what we can sense?"

EAGLEMAN

We shrunk the Neosensory vest down into a wristband with four tiny motors on the band (called the Neosensory Buzz) and this is now on wrists around the world. I'm working on many experiments now, such as passing information from infrared and ultraviolet light around me. I'm interested in the fact that most of the light which surrounds us is totally invisible to us. At least, until now.

I'm also exploring the idea of detecting other people's physiology. So imagine you're wearing a smartwatch, and I can feel what's happening with your heart rate and galvanic skin response and heart rate variability. Over the internet, I'm feeling that, and so if, after we're off this Zoom call, I'm thinking, "Oh man, Daniel feels a little stressed out," I can check back and see if everything is okay.

Anyway, there's all kinds of things we're trying. Electromagnetic fields. Picking up the social context of someone else's speech, so that if you're autistic, for example, you can read someone's social cues.

KWAN

That's wild. You're turning a smartwatch into a sensory hive mind experience.

SCHEINERT

So one day, your doctor might just wear 50 watches for all his patients and be able to feel, "There's something wrong with Daniel?"

[Laughter]

EAGLEMAN

That's right.

KWAN

That's fantastic. In your eyes, what's the end game of this technology? What does society look like once it's everywhere?

EAGLEMAN

This is a general platform for passing on any data stream you want. We've had to inherit all these other sensory devices through a long road of evolution, but now this can serve as [an additional sensor for] anything you want to pick up on. I think where this might play out in society is, for example, if I'm picking up on infrared information, and you're picking up on stock market information or whatever, we might end up having a slightly tougher time communicating. I can't understand what you're feeling, and you can't understand what I'm feeling. Nonetheless, it becomes like one of those superhero movies where everyone's got their own superpower, which is, by the way, not that different from how it is now. I mean, people pick up on different kinds of things now. If you're already a dermatologist or an architect or a lawyer, everyone's hearing different things out of the conversation because of their expertise, right?

SCHEINERT

That was one of the wilder concepts I took away from *Incognito*,[6] that when we say that two people have a different point of view, it can be literally true brain-wise, because our brain is fabricating so much of what we think we're observing. I have a different past, so I have a different brain, which makes different assumptions. So Dan and I, for example, we're observing literally different worlds. That's pretty scary, but also makes a lot of sense.

5. David helped start a company called Neosensory, based on the idea that the brain can take any sensory information and learn to interpret it as sight, sound, or as a sixth, seventh, or eighth sense. They are creating vibrating technology that functions as affordable additional input devices for your brain… Vibrators for your brain! (He's going to hate that we made up that slogan, isn't he?)

KWAN

But you also talk about how we're not only processing our own worldview, but we're sharing it with others. It's like taking a picture: I take a picture, and I share it with someone else. Now we have a shared experience. But on a more meta level, right now, one of the biggest problems I think we're facing is the problem of processing information. The difference between a future where we succeed and the future where humans just kind of completely collapse is whether or not we find a solution to the problem of the amount of information we have and how to process it properly and effectively among all of us, as the network grows and the complexity of ideas grows. There are so many more moments where breakdown can happen. And that's where we're at right now. Do you believe that this gadget that you're creating can also help us to understand each other? Get to this better, deeper, physiological version of empathy?

EAGLEMAN

Here's what I would say. I think that the advent of the printing press was the first important step that gave people an opportunity to step into each other's shoes in a way that had never been done before. And literature has just gotten better and better at that. You can absorb literature from around the world and from different perspectives. Of course, movies also do that all the time in ways that are very easy to absorb in two hours. You get to step into other people's universes, or multiple universes, as the case may be.

As far as how to consume the information—I mean, today we have the opportunity to absorb all these different viewpoints. It feels like kids have better internet literacy than people our age did growing up, in terms of understanding that one person's opinion isn't the truth. And so as long as we train our kids how to "read the internet," well, I'm not too worried about that.

Let me just explain why. One of my other books is called *The Safety Net*, and it examines what the internet means for civilizations on a 10,000 year timescale. I make the argument that the internet makes government censorship impossible now. It takes away that power from authoritarian governments, because now citizens can just click around for other points of view. Governments can't control the news as they did last century—say in the Soviet Union, communist China, or Nazi Germany. That's great news for us. We get the opportunity to dine on a broad diet of opinions, and we simply need to properly educate our children to understand the difference between points of view and truths. This opening of the gates is the perhaps the biggest benefit of the internet.

KWAN

That's hopeful. I like that.

SCHEINERT

We have this forever-going debate about internet pessimism and internet optimism, and you have written some of the more internet-optimistic stuff that I've read. Do you get a lot of pushback on that?

EAGLEMAN

I don't think so. Whenever I've been asked this question at a public talk and I give my answer, I find that a lot of people feel relieved, because there's been this story from the beginning of, "the internet's making us dumber." And I think it's just the opposite. I'm so enthusiastic. I mean, my nine-year-old kid is so fucking smart precisely because he constantly watches TED videos and BrainPOP videos. He just knows so much stuff that I had no clue about when I was nine years old.

SCHEINERT

It's tough sometimes to know if the terrifying misinformation stories we're reading about are worse than ever, or we can just see them better than ever.

EAGLEMAN

There's absolutely nothing new about misinformation. It used to be so much worse, because there was one authority that told you what was up: the Nazi government, or Pol Pot in Cambodia, or whoever. Now it's easy to circumvent that; it's easy to find other points of view.

---

6. In *Incognito: The Secret Lives of the Brain*, Eagleman collects some of the wildest stories and studies about the subconscious, ultimately illustrating how little control we actually have over our behavior, putting everything we know about free will, culpability, and our modern legal system into question. It's a humbling read!

KWAN
One thing that we've been obsessed with lately is how the attention economy makes it impossible for smart people's ideas to make it into the mainstream and break through the surface to become part of the lexicon. So as a smart person, if you were given $300 million to make a movie right now, knowing that you would have the attention of a lot of people, what story would you want to tell? What is the David Eagleman blockbuster?

EAGLEMAN
I can answer that because I've already written a few screenplays [laughs]. One of them is based off a story in *Sum*. The idea is that we can feel so certain about things, but that, fundamentally, there are things you don't know, and we constantly find our internal models breaking. For example, in any movie you have some sort of plot, and it's super easy, as you guys well know, for the storyteller to lead the reader down a garden path, even if the storyteller is intending to have a surprise, the twist, whatever. The reason it's so easy is because you just need to give a few facts and clues, and then the audience's internal models are like, "Oh yeah, I got it. I know the truth. I know exactly where this is going." Then you can surprise them by going the other way. That's why it's shockingly easy to write a mystery novel.

[Laughter]

KWAN
Shots fired!

EAGLEMAN
Only because it's terrifically easy to make assumptions about [how people will fill in the blanks]. Anyway, my blockbuster is a movie about that—about thinking at all moments that you know what's going on, but there are deepening mysteries that illustrate how much we don't know.

KWAN
So you're saying these narrative itches are already inside the human brain and that the storyteller then manipulates them. We hijack it—that's what storytelling is. I think movies that acknowledge and actually play with that are really interesting, especially now, because I think everyone should be very much aware of [that manipulation].

SCHEINERT
Thinking about how huge the unknown aspects of the world are and how this universe might just be a bit of a neurological illusion—whether it's the measly three dimensions we can observe, the crazy assumptions our brains are making, or the endless number of universes we're not observing—we wanted to ask what theories and beliefs that we hold dearly today do you think future generations might laugh at us for having? Any hot takes?

EAGLEMAN
What is our generation going to get damned for by future generations that we have no idea about? Maybe something like eating animals—many of us eat animals now and we don't think twice about it, but almost certainly in a hundred years, we're gonna all be crucified for that. Certainly, all of the energy consumption will be demonized by future generations.

SCHEINERT
Yeah. Maybe in the future, hyper-intelligent cattle will laugh at this naive interview that some cruel monkeys conducted and decided to print in a book.[7]

KWAN
What a confusing last thing to say in the interview. Well, thanks for talking about the unknown with us David. We're sorry in advance for future generations' inevitable backlash.

7. We added these jokes to the interview, so it wouldn't abruptly end with eating animals. The real interview ended with us talking about how we had gone over our allotted interview time and how hard it was going to be for someone to edit all this down. Thanks so much to the folks who helped edit all this down.

**At Age 28, Chilean Astronomer Maritza Soto
Has Already Discovered Three Planets**
by Vincent Toro

*This poem takes its title from the headline of an article
published by* Remezcla *on Sept. 21st, 2018.*

Haloed by the glow of the multiverse swirling
above La Silla Observatory, your pyrex eye
spotted an orb three times the mass of Jupiter.
    All these lenses leering at the heavens,
    and yet it was you who identified
    HD110014C. You were reluctant to call
 it discovery, perhaps because you know
 all too well what poisons gush forth
 from that word. Or maybe you suspect
      you are not the first because you
      know there is no such thing
      as firsts. Still, you did what no
gringo ever could: you made another world
visible to nosotrxs. Perchance it was HD110014C
that actually recognized you long before your
    spectroscopic lens detected her.
    It might even be that she had already
    decided to entrust you with making
      her presence known to our kind.
      After all, you proved yourself more
      than worthy of such responsibility
when you said your
finding was "not
exceptional," annihilating
    the misguided western patriarchal notion
    of greatness too many others have used
    to boost themselves since 1492.
      You even confessed your introduction
      to HD110014C
      was entirely an accident,
 a courageous admission that eclipses
 the bumbling arrogance of every Columbus,
 every Cortez, every Pizarro. From 300 million
    light years away you glimpsed
    another possibility, then befriended
    two more exoplanets before

your 28th year around
our lilliputian sun. You,
sprung from a country
             crystallized in its mourning
             of the disappeared,
             met a glorious
                          dawn and flash
                          fused to emerge
                          as one
                 woman search party.
                 Maestra Maritza, I know
                 this goes against all
scientific wisdom, but I can't help but theorize
that these three interstellar marvels you've pulled
into our orbit have become a new home for those
             that collapsed into the event horizon
             of imperial cruelty. I like to suppose
             that our gente were never erased
but rather beamed to a star system
that does not regard them as merely tool
or trinket, a galaxy where their dreams
                          are as important as those
                          who dwell in some imaginary
                          North. Could it be, Maritza,
that what you scoped out there among
the shimmering Allness was in fact
a reunion pachanga thrown on the gold
             dust rings of a wandering star where discovery
             is not a sword of Damocles but instead a feathered
             reentry path for those who have been missing us.

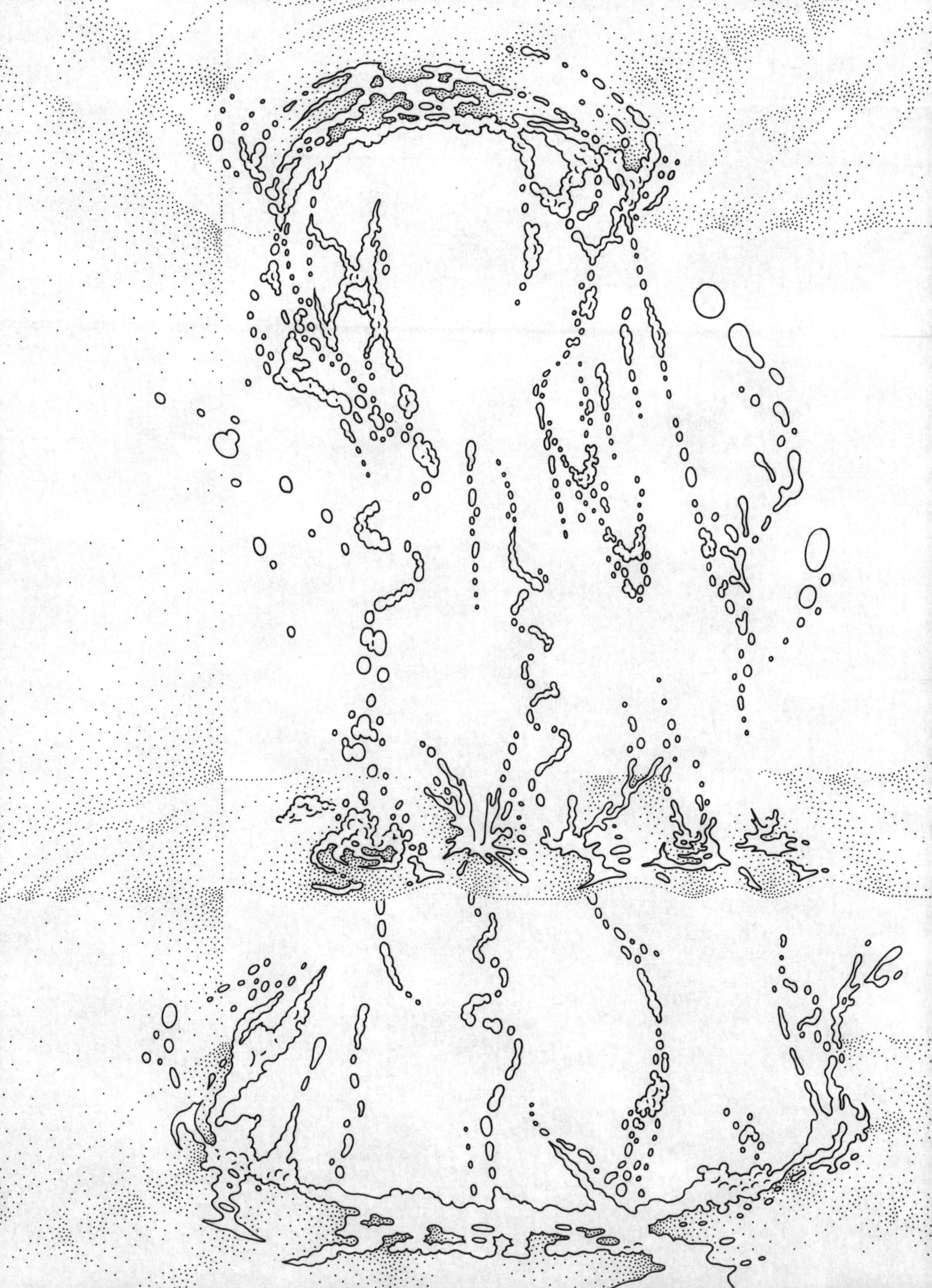

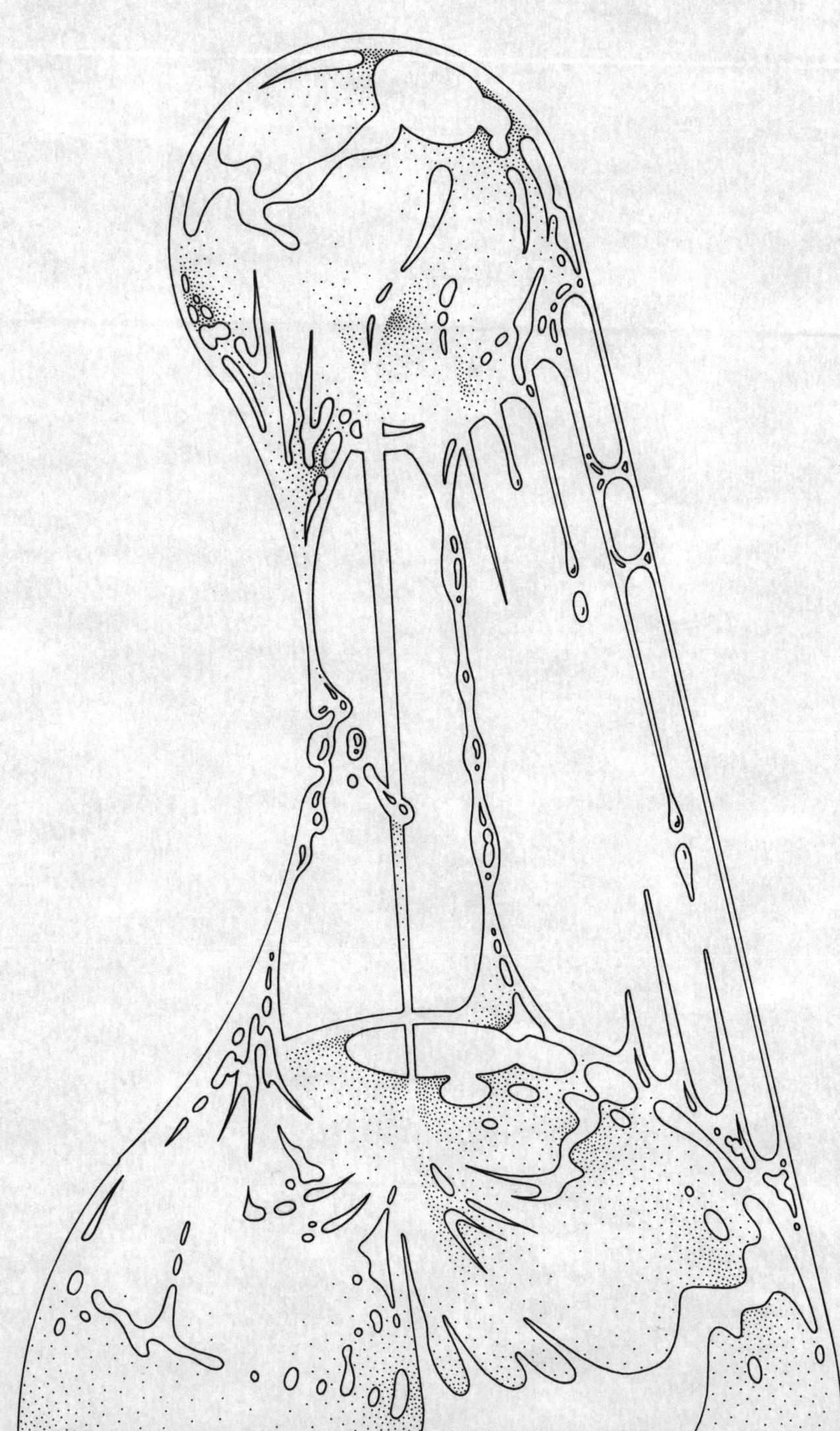

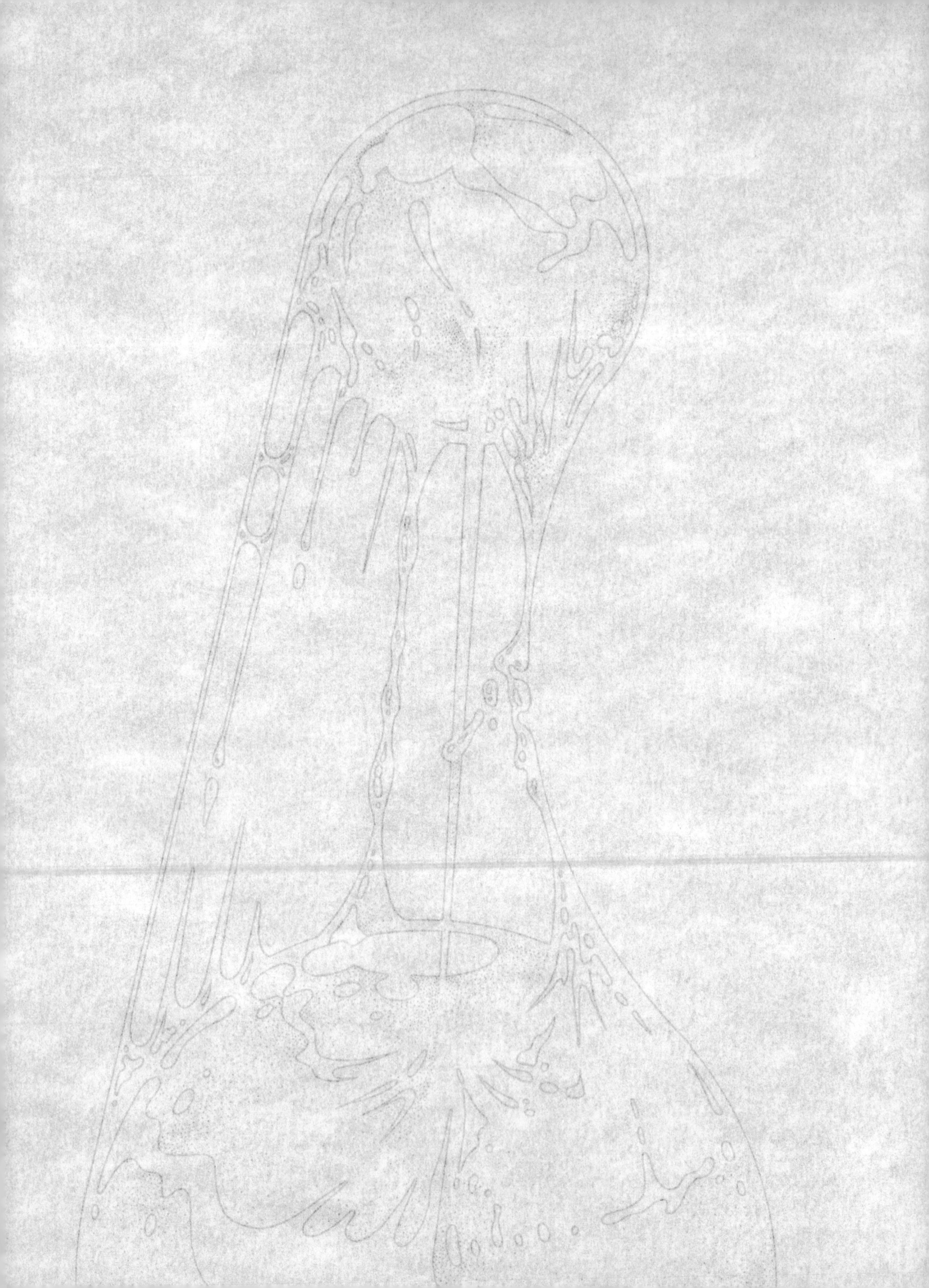

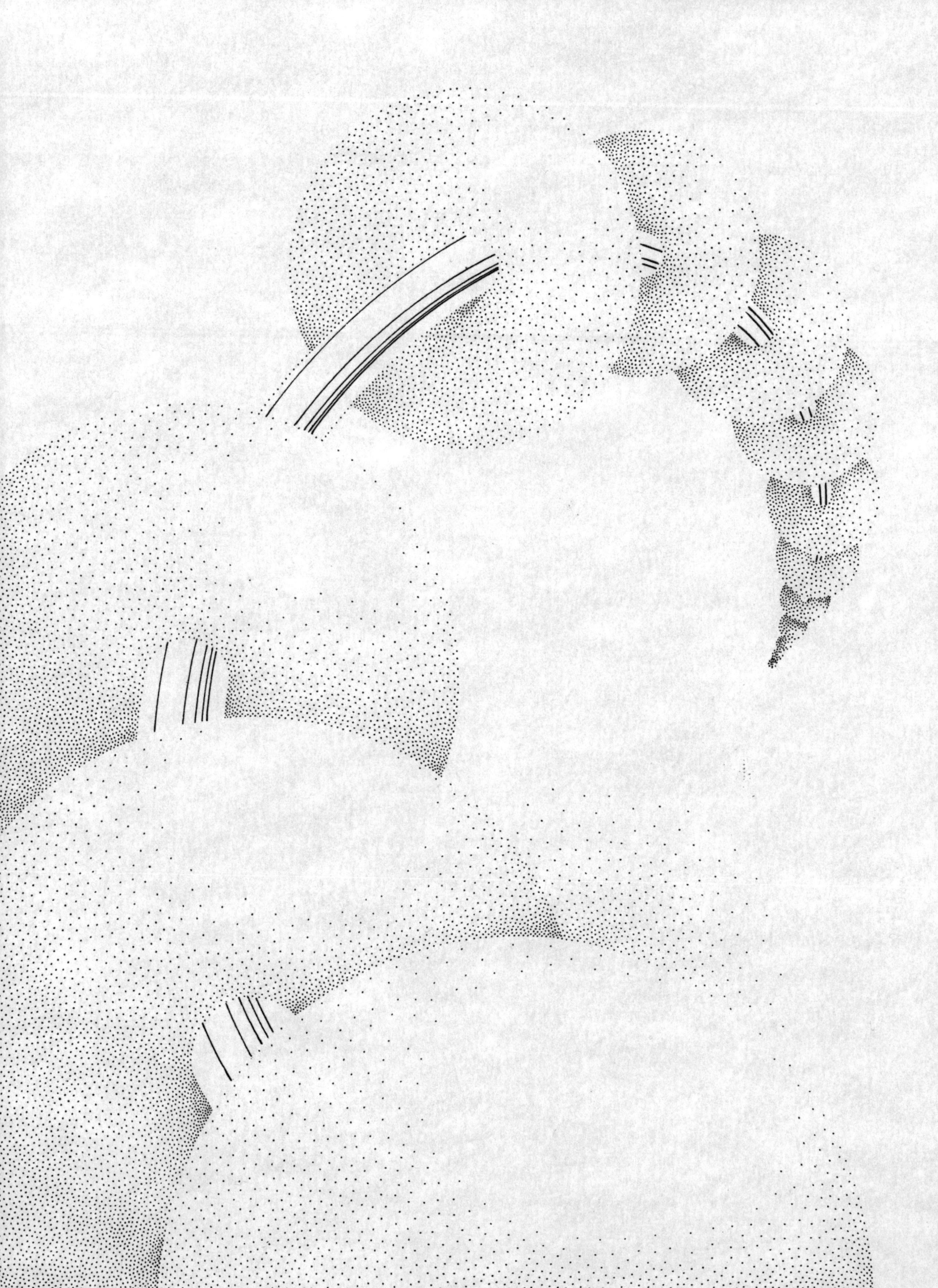

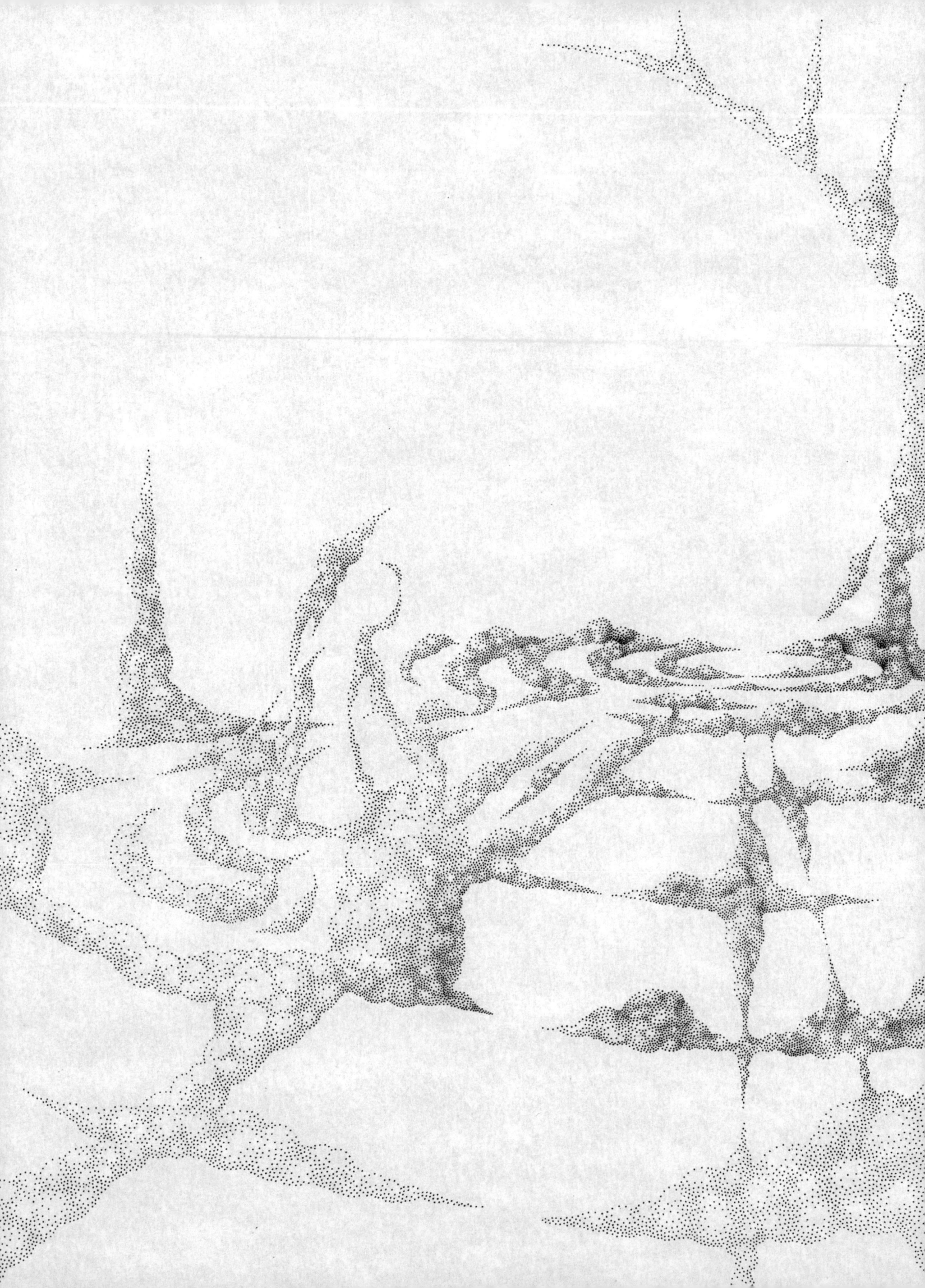

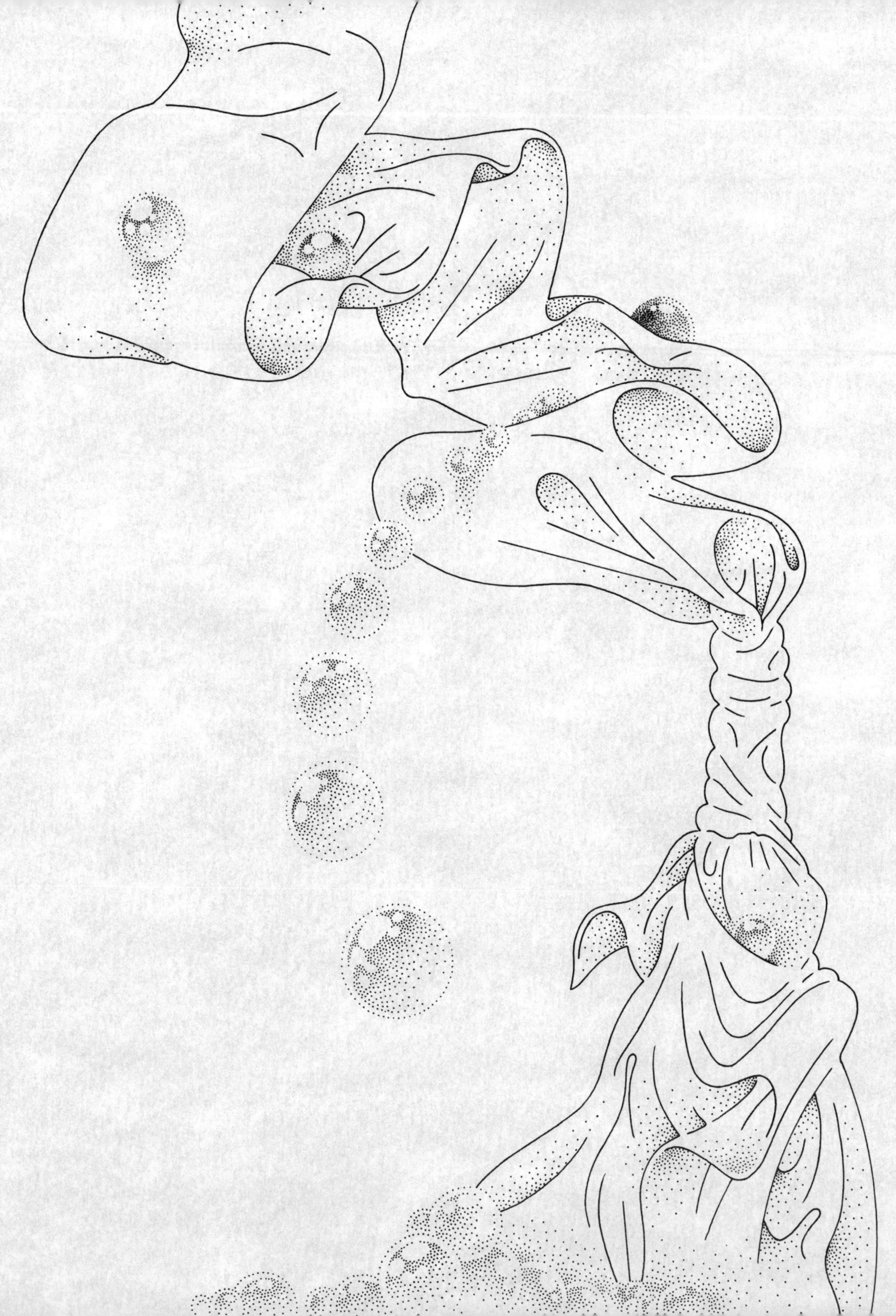

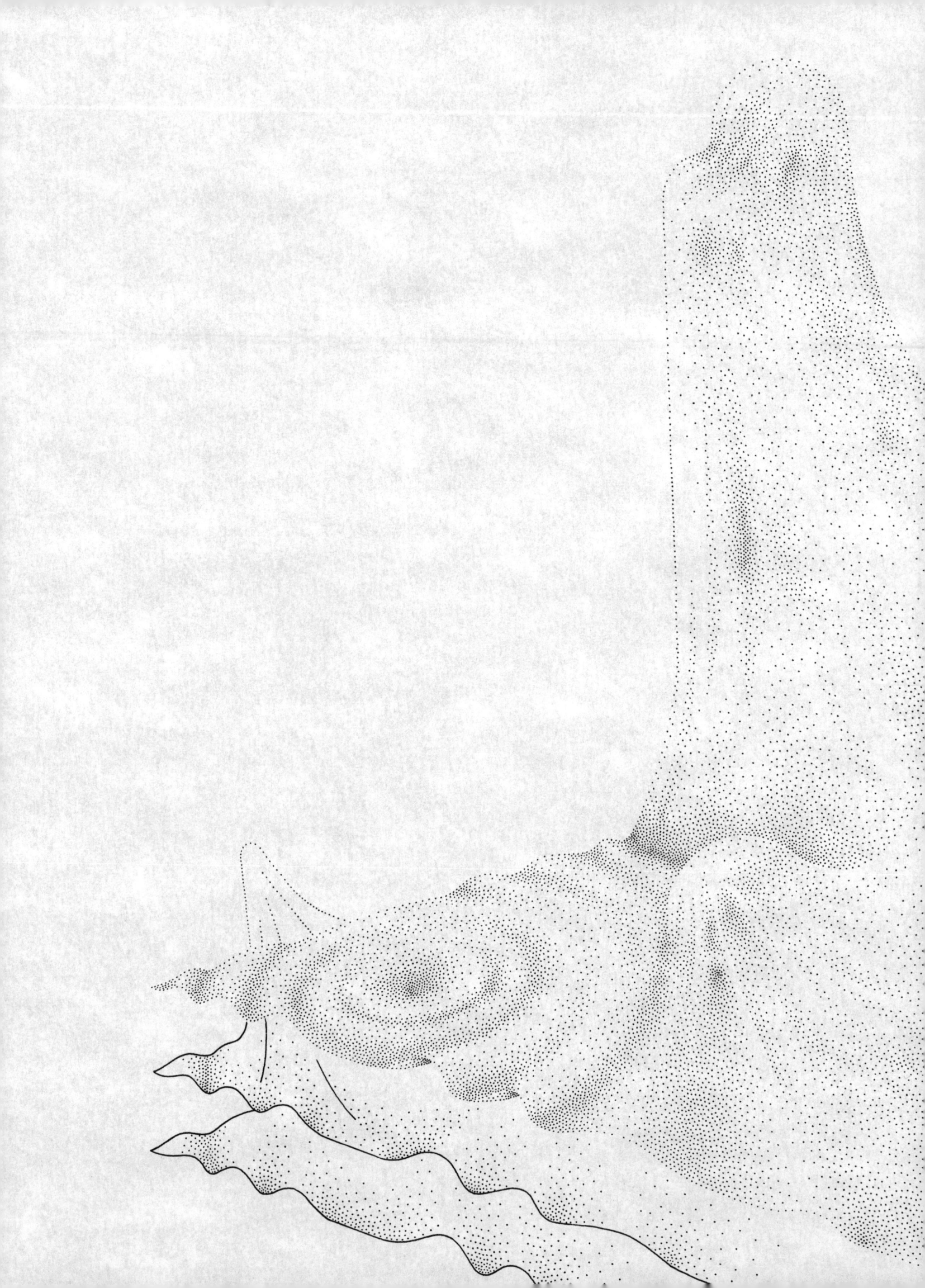

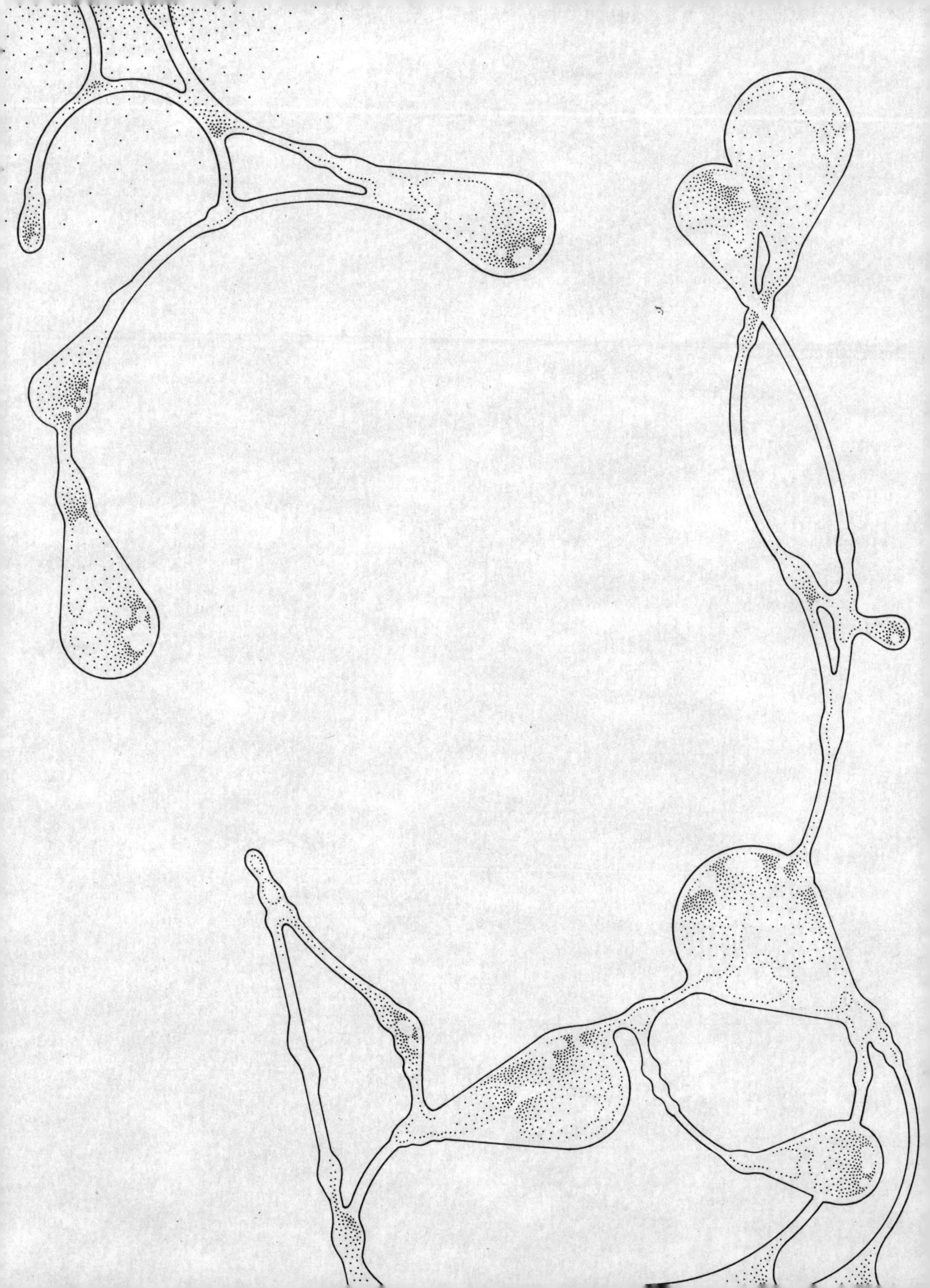

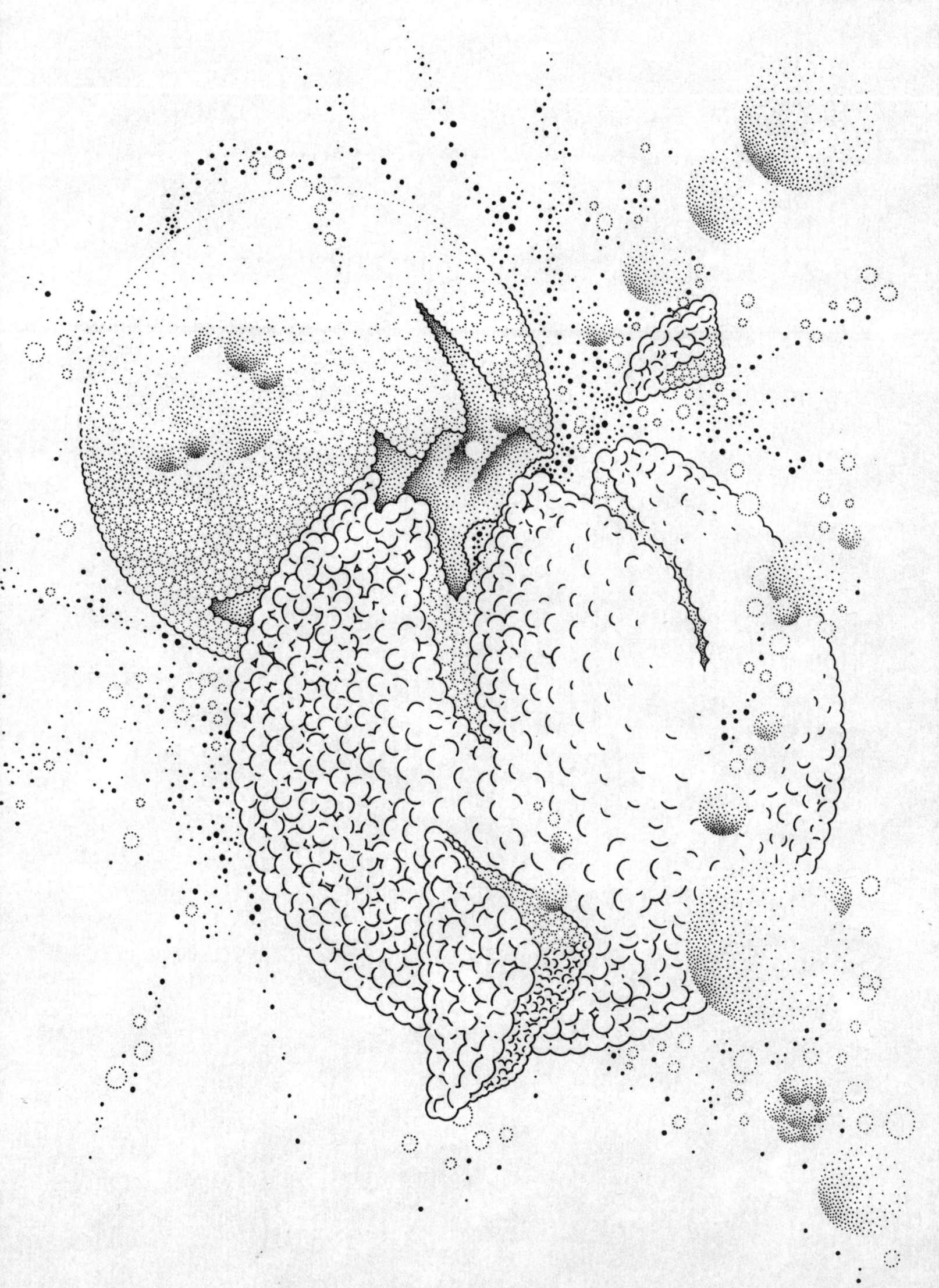

Good Bones
by Maggie Smith

Life is short, though I keep this from my children.
Life is short, and I've shortened mine
in a thousand delicious, ill-advised ways,
a thousand deliciously ill-advised ways
I'll keep from my children. The world is at least
fifty percent terrible, and that's a conservative
estimate, though I keep this from my children.
For every bird, there is a stone thrown at a bird.
For every loved child, a child broken, bagged,
sunk in a lake. Life is short and the world
is at least half terrible, and for every kind
stranger, there is one who would break you,
though I keep this from my children. I am trying
to sell them the world. Any decent realtor,
walking you through a real shithole, chirps on
about good bones: This place could be beautiful,
right? You could make this place beautiful.

Cyclical

The question of how the universe will end turns out to be the same as asking in which way the universe will recreate its birth.

If current cosmic trends continue, space will gradually expand faster and faster until, trillions of years from now, it will blast apart exponentially, much like most cosmologists think happened during cosmic inflation at the start of the Big Bang. But it's also conceivable that space will eventually slow its expansion, then stop, then start to contract, ultimately shrinking back to the point from which it came.

If the universe expands and then contracts, it will experience either a Big Crunch or a Big Bounce. If it bounces, then perhaps the universe is like a lung, expanding and contracting over and over again.

Paul Steinhardt, one of the pioneers of eternal inflation in the early '80s, turned against inflation because of its attendant multiverse of bubble universes; he now supports the cyclical model. Steinhardt argues that the inflationary multiverse is unscientific, because, in predicting everything—every imaginable bubble—it predicts nothing. Steinhardt prefers to think our universe is self-contained, a lone lung (or maybe a pair of lungs, as we shall see).

He and his collaborators contend that, in a cyclical universe, the contraction phase can do the work of cosmic inflation, rendering space smooth and flat. Other researchers speculate that, as the universe shrinks, the arrow of time would reverse, and the entropy, or disorder, of the universe—which we see as rising—must fall. That means the world would become ever-more orderly, and meanwhile, energy would grow more concentrated. You'd see pieces of shattered glass magically rejoin, which would be nice. On the other hand, fire would suck all the heat from your body, which would be not so nice.

Researchers have also been studying whether and how it's possible for a universe to bounce, going from contracting to expanding. In some models, the bounce occurs when our brane collides with another parallel brane; the pair of branes might bounce off each other repeatedly forever.

Investigations of cyclical models continue in the shadow of research on cosmic inflation, which has more adherents. Whereas inflation yields a multiverse of bubbles, a cyclical universe undergoing eternal rebirths would play out infinite stories sequentially as a multiverse of another name.

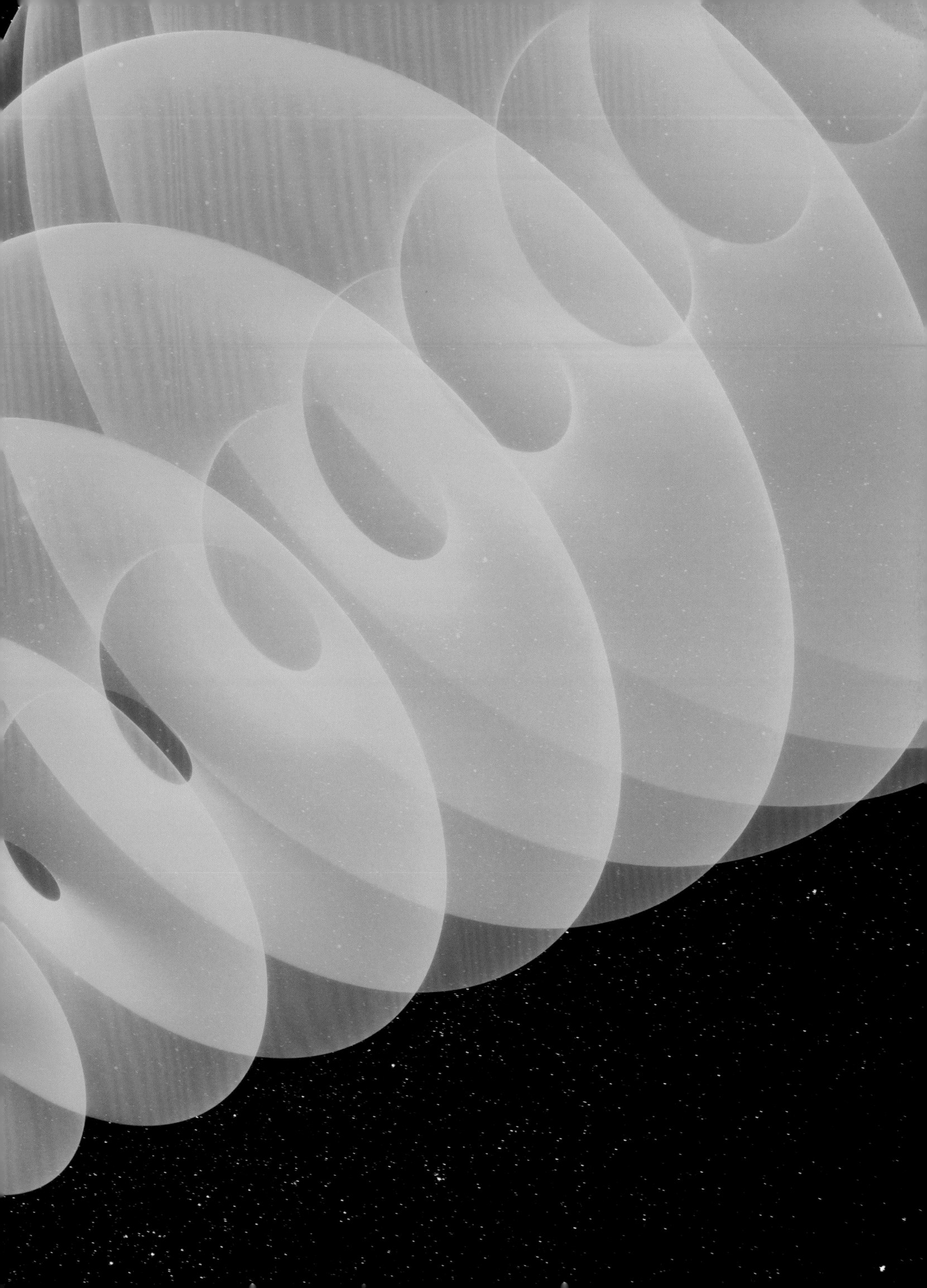

Paradise
by Eileen Myles

which
I never know
how to
spell it's
with me always

Haiku
by Nancy Huang

wait for me at the
ampersand; i am still here,
looking for the end

Reversal

by David Eagleman

There is no afterlife, but that doesn't mean we don't get to live a second time.
At some point, the expansion of the universe will slow down, stop, and begin to contract, and at that moment the arrow of time will reverse. Everything that happened on the way out will happen again, but backward. In this way, our life neither dies nor disintegrates, but rewinds.

Cyclical

In this reverse life, you are born of the ground. At funeral ceremonies, we dig you up from the earth and transport you grandly to the mortuary, where the birth makeup is removed. You then are taken to the hospital, where, surrounded by doctors, you open your eyes for the first time. In your daily life, broken vases reassemble, meltwater freezes into snowmen, broken hearts find love, rivers flow uphill. Marriages reride rocky roads and eventually end in erotic dating. The pleasures of a lifetime of intercourse are relived, culminating in kisses instead of sleep. Bearded men become smooth-faced children who are sent to schools to gently strip away the original sins of knowledge; reading, writing, and mathematics are expunged. After this diseducation, graduates shrink and crawl and lose their teeth, achieving the purity of the highest state of the infant. On their last day, howling because it is the end of their lives, babies climb back into the wombs of their mothers, who eventually shrink and climb back into the wombs of their mothers, and so on, like concentric Russian dolls.

In this reverse life, you have blissful expectations about what will come next as you experience your story backward. At the moment of reversal, you are genuinely happy, for while life must be lived forward the first time, you suspect it will really be understood only upon replay.

But you have a painful surprise in store. You discover that your memory has spent a lifetime manufacturing small myths to keep your life story consistent with who you thought you were. You have committed to a coherent narrative, misremembering little details and decisions and sequences of events. On the way back, the cloth of that story line unravels. Reversing through the corridor of your life, you are battered and bruised in the collisions between reminiscence and reality. By the time you enter the womb again, you understand as little about yourself as you did your first time here.

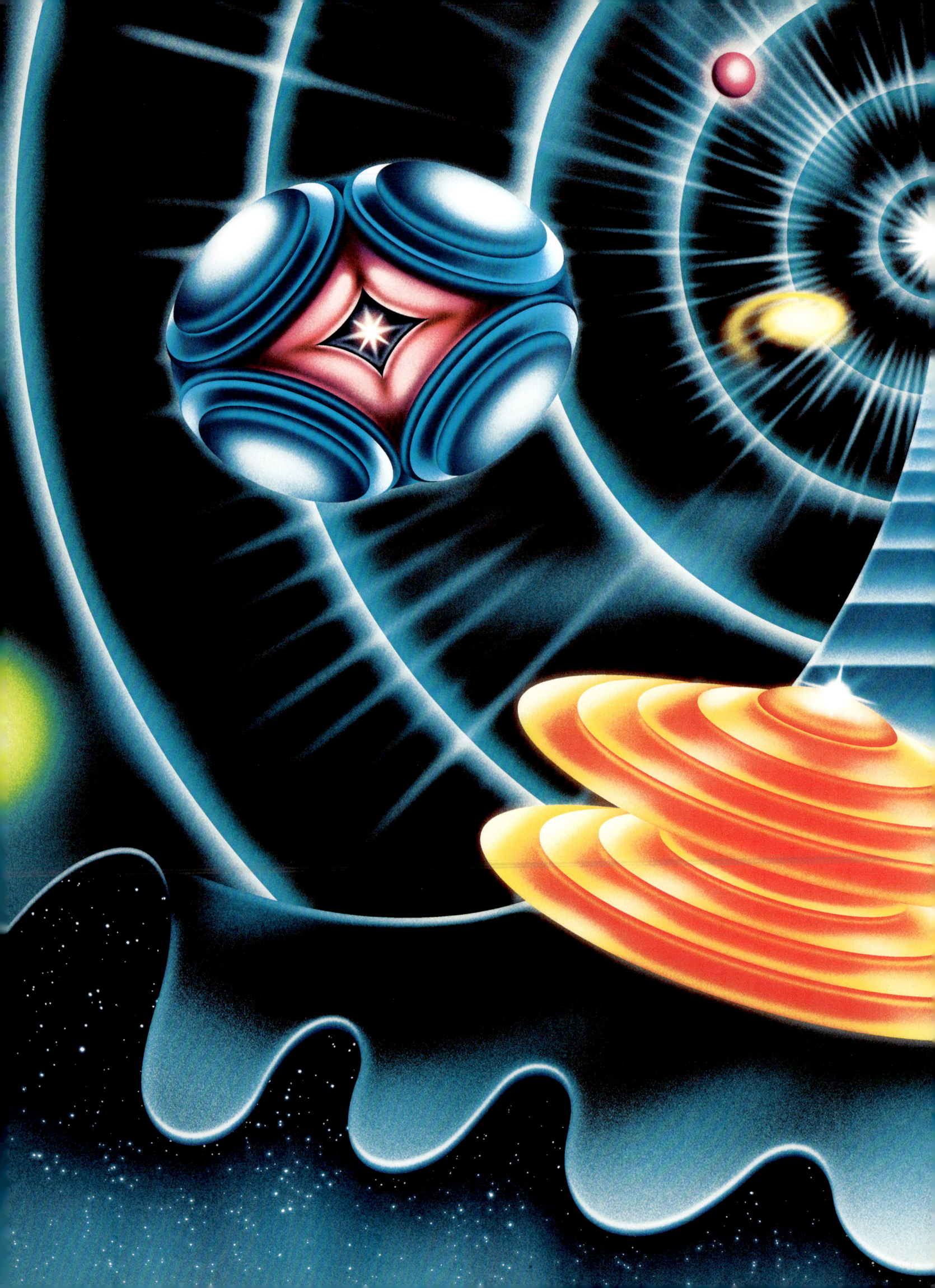

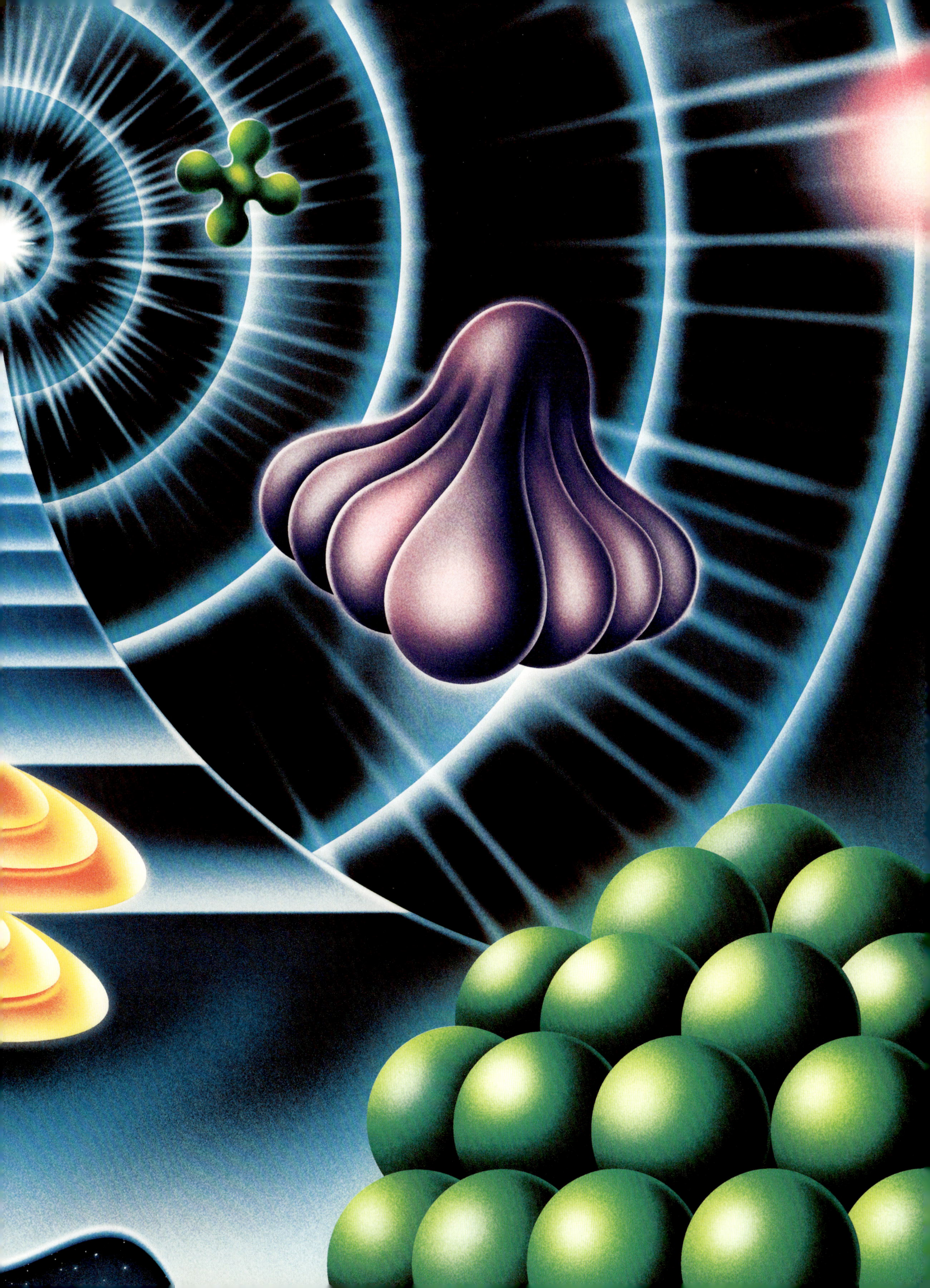

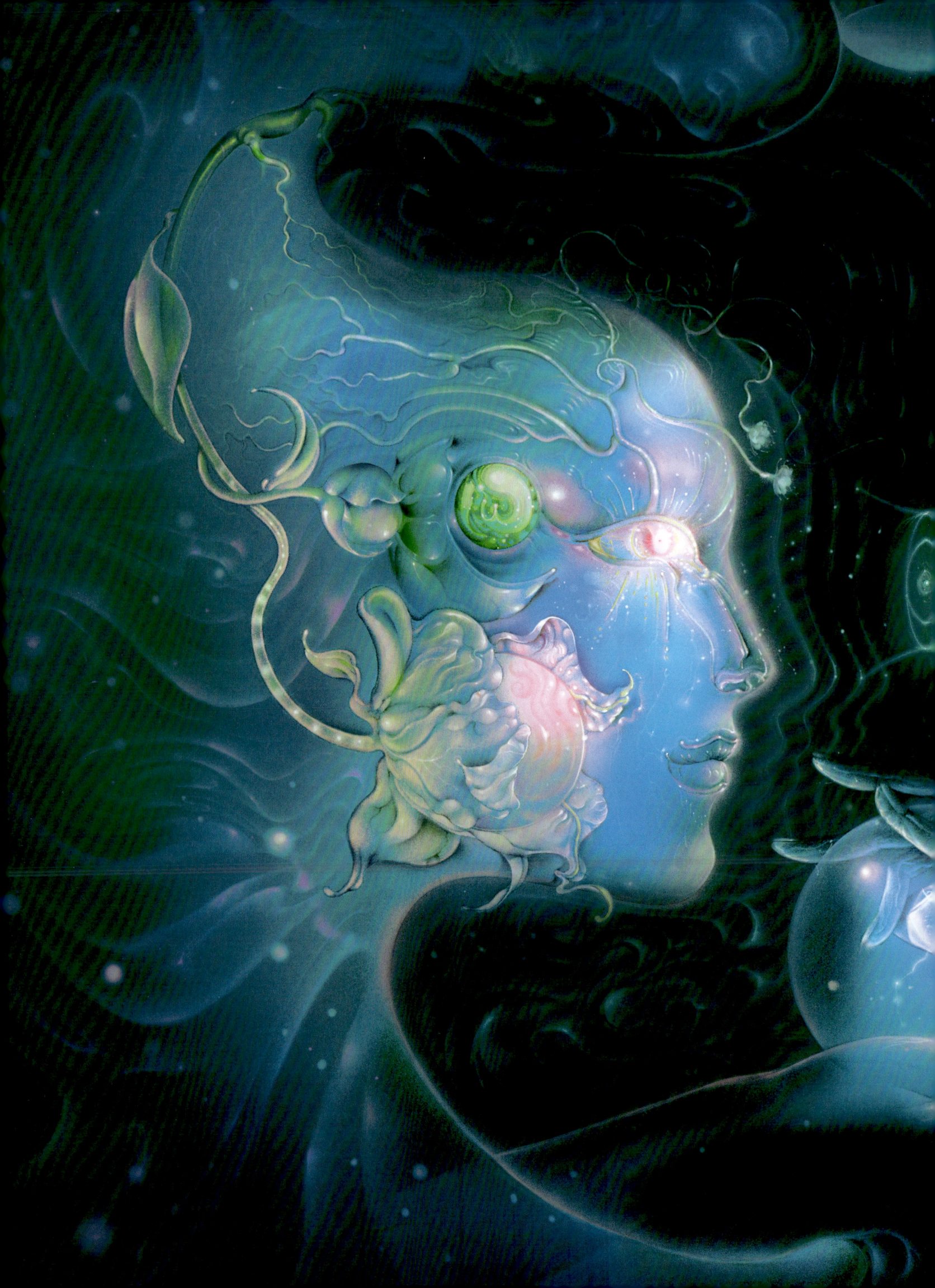

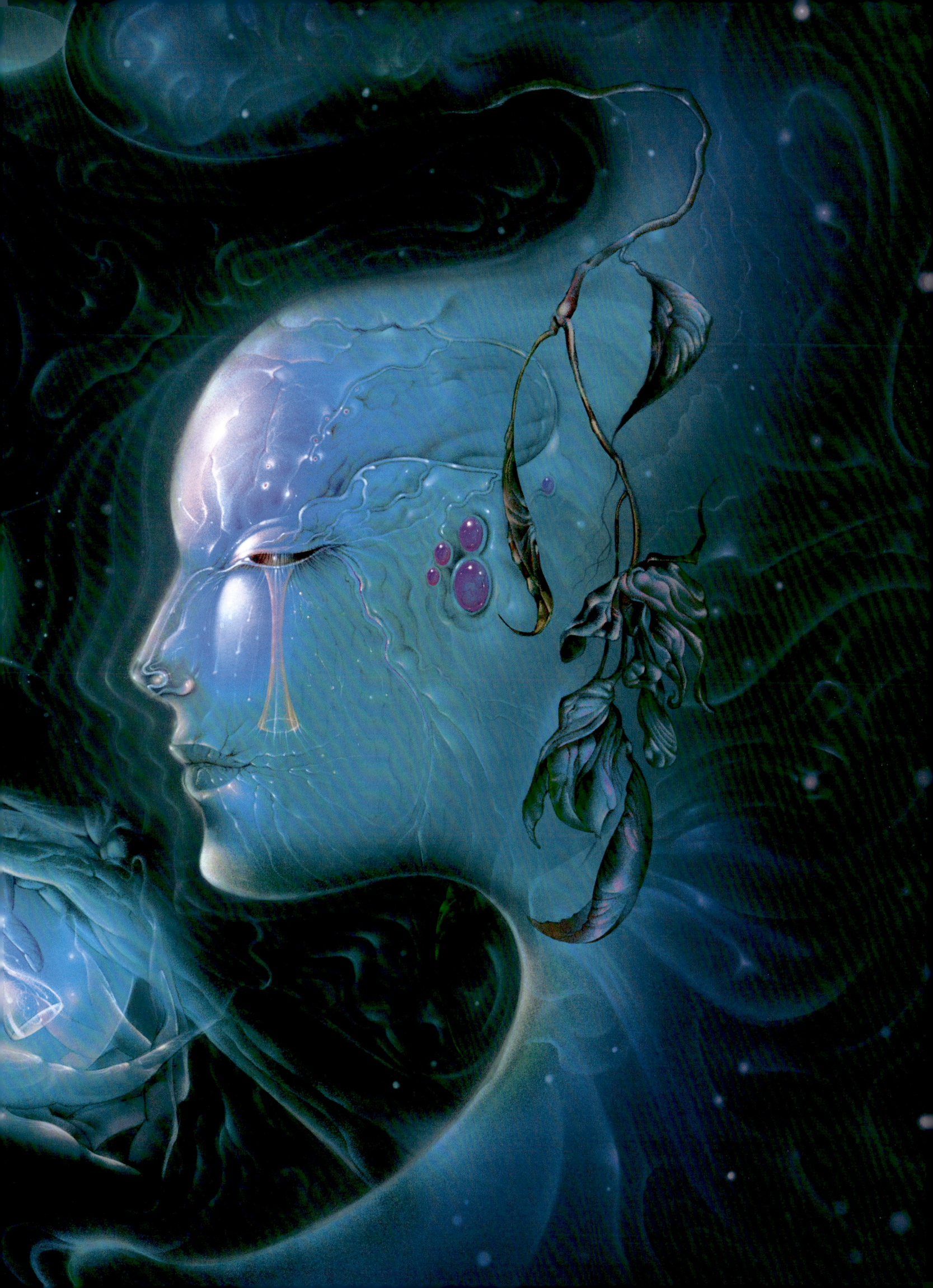

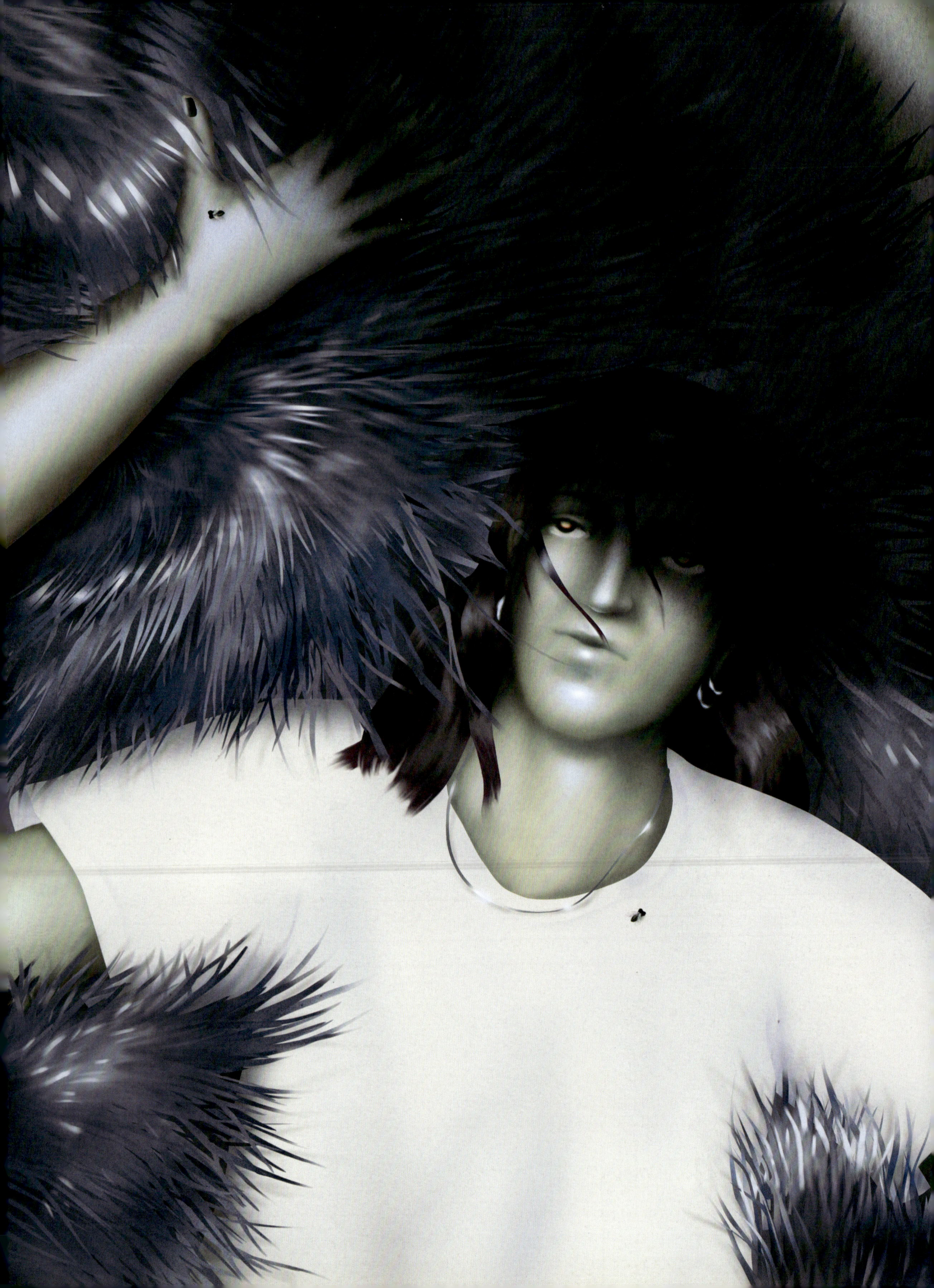

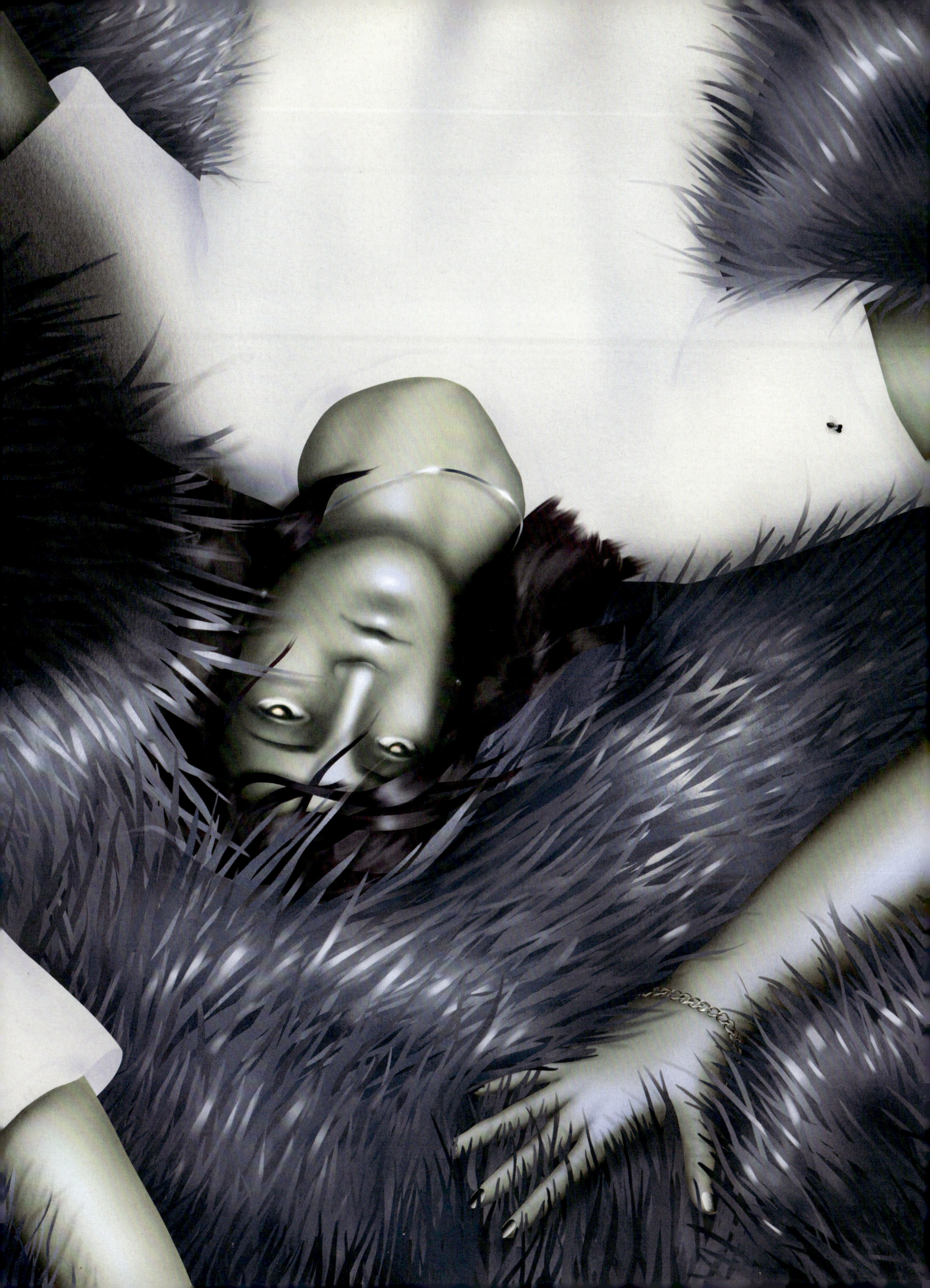

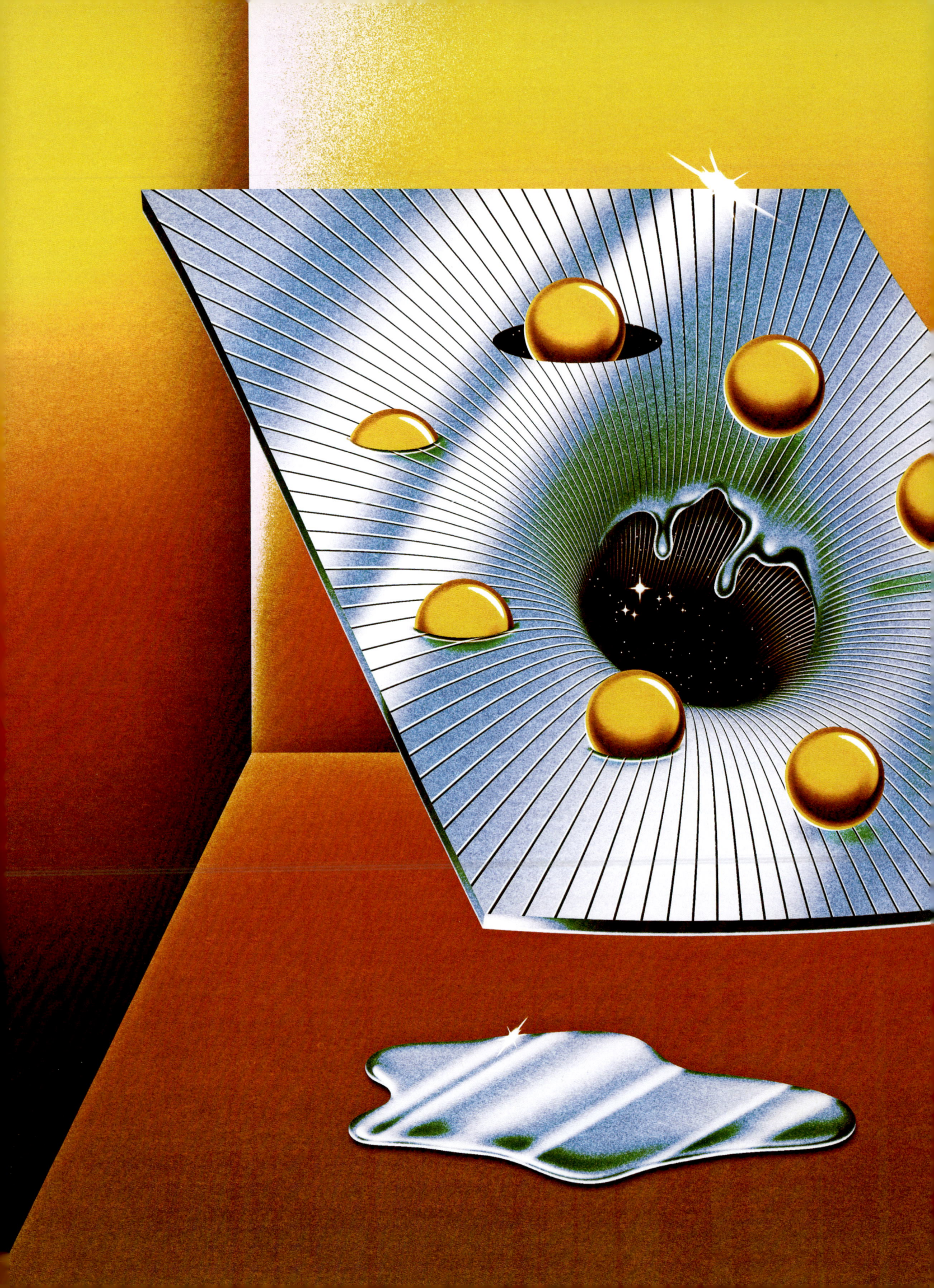

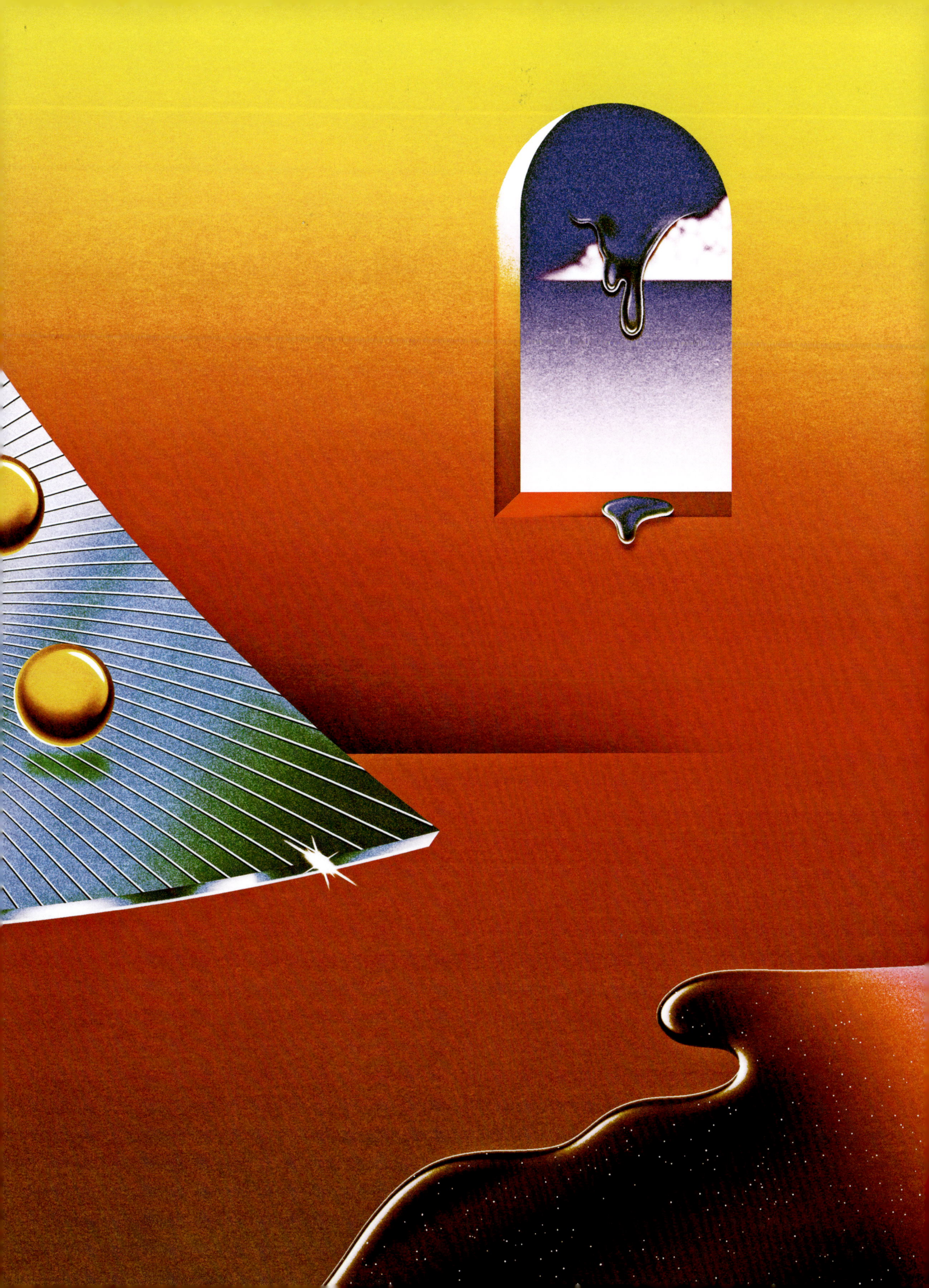

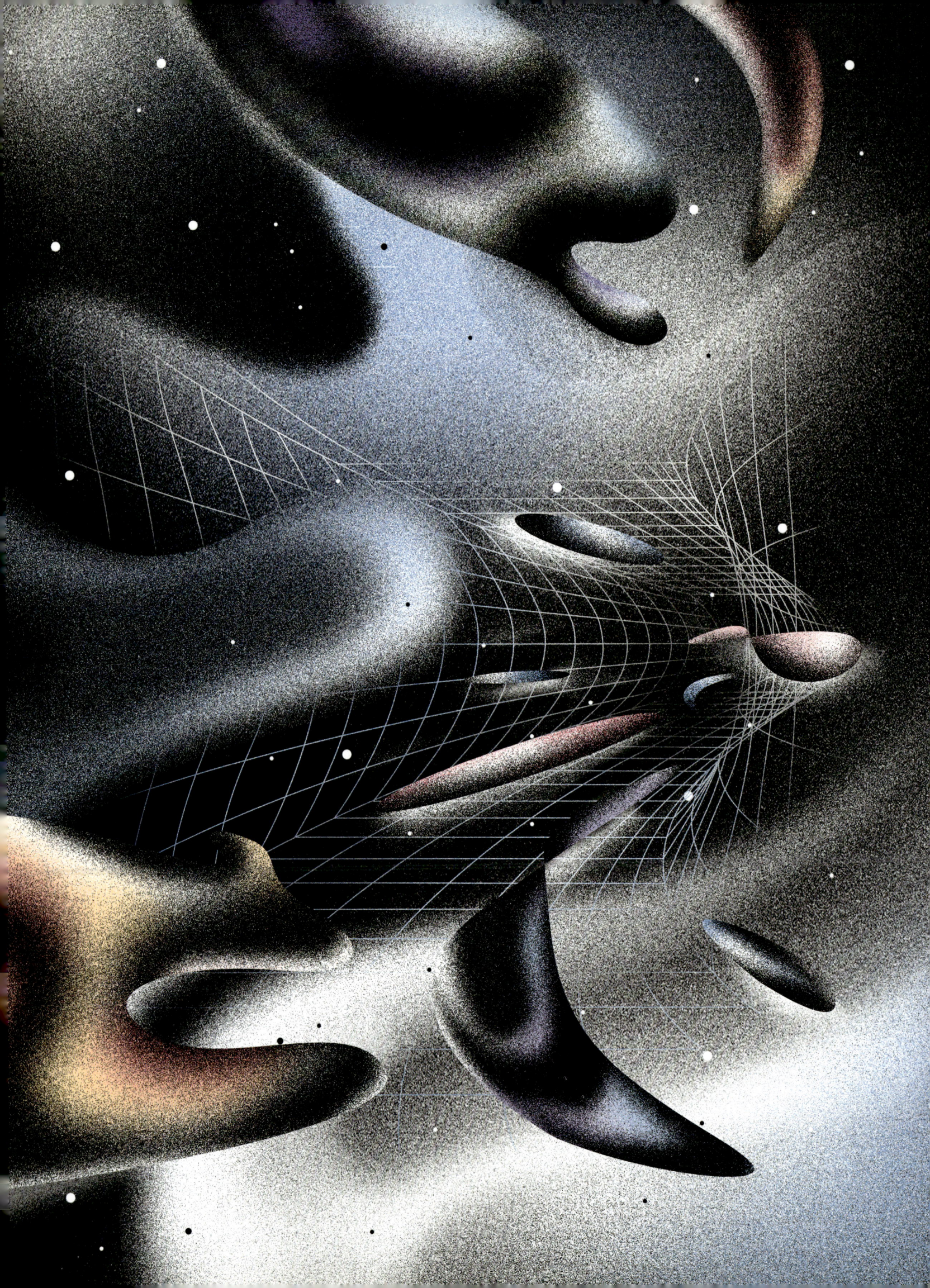

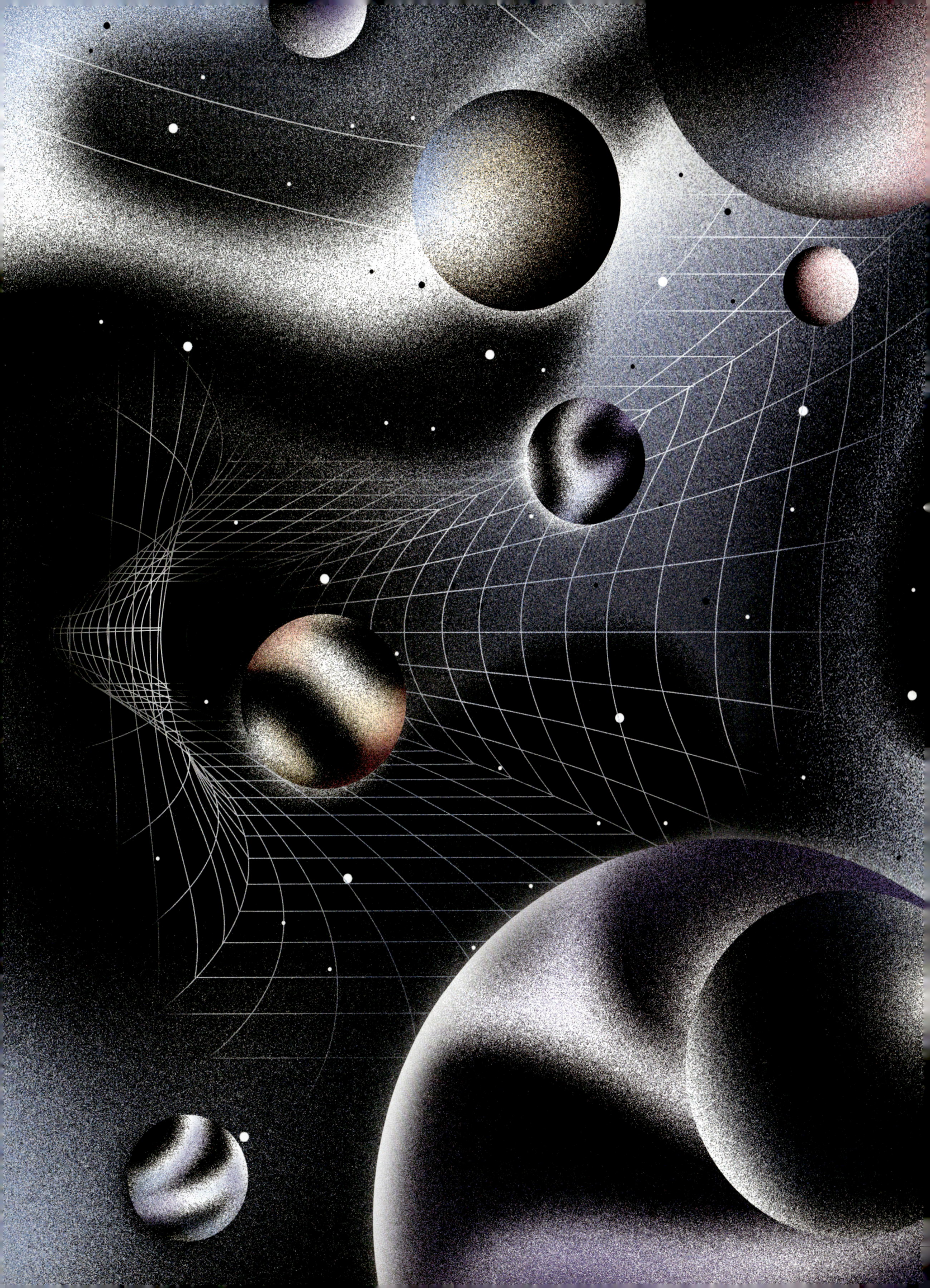

WE AS ORGANISM

A lecture by Alan Watts

I wonder if it's ever struck you how curious a thing it is that most of the things that we experience we regard as things that happen to us, which we ourselves do not originate, which are events expressing some sort of power or activity that is external to ourselves. And if you consider that, you realize that what you mean by "yourself" is rather narrowly circumscribed. Even events that go on in our own bodies are put in the category of things that happen to us in the same way as things that go on in the world outside our skins. If there's a thunderstorm or an earthquake—well, it happens to you; you're not responsible for it. But so, in the same way, when you have hiccups you didn't plan on it. If you have belly rumbles, you had no intention of doing it. And as for the catastrophic act of getting born... well, you had nothing to do with that. And you can spend all your life blaming your parents for putting you in the situation in which you find yourself.

This way of looking at the world in this sort of passive mood is similar to the way in which, as Westerners, we have been accustomed to look at human existence as a precarious event in a cosmos that, on the whole, is depicted as being completely unsympathetic and alien to our existence. In other words, if you're reared with a 20th-century common sense (which is based on the philosophy of science of the 19th Century with its rejection of Christianity and Judaism), you regard yourself as an accident—a biological accident—in a stupid universe that is mechanical but has no feelings—no finer feelings. A vast, pointless gyration of radioactive rocks and gas in which you happen to occur.

Of course, if you don't have that point of view and you are more traditional, you look upon yourself as a child of God and therefore under authority. In other words, there's a big boss on top of all this who allowed you, at his pleasure, to deign to have the disgusting effrontery to exist, and you better watch your Ps and Qs, because that Almighty is looking after you with the attitude of "this is going to hurt me more than it's going to hurt you."

And when you look at the world in that image—or in the other image that it's a stupid mechanism—either point of view you take, you don't really belong. You're not really part of all this. And I could use a stronger word than "part"—only, we don't have it in English. We have to say something like "connected with it," "essential to it." Or, to put it in the strongest possible way, it is alien to Western thought to conceive that the external world—which is defined as something that happens to you, and your body itself is something that you got caught up with—it is quite alien to our thought to consider all that as you, yourself. Because, you see, we have such a myopic view of what one's self is. It's as if, in other words, we selected how much experience is to be regarded as "me," as if you focused your attention on certain restricted areas of the whole panorama of things that you experience and say "I will take sides with that much of it."

Now, we come here—right at the start—to an extremely important principle, which is the different points of view you get when you change your level of magnification. That is to say, you can look at something with a microscope and see it a certain way; you can look at it with a naked eye and see it in a certain way; you look at it with a telescope and you see it in another way. Now, which level of magnification is the correct one? Well, obviously, they're all correct, but they're just different points of view. You can, for example, look at a newspaper photograph under a magnifying glass and where, with the naked eye, you will see a human face, with a magnifying glass you will just see a profusion of dots rather meaninglessly scattered. But as you stand away from that collection of dots, which all seem to be separate and apart from each other, they suddenly arrange themselves into a pattern. And you see that these individual dots add up to some kind of sense.

Now, you'll see at once, from this illustration, that maybe you—when you take a myopic view of yourself, as most of us do—may add up to some kind of sense that is not apparent to you in your ordinary consciousness. What is, in other words, conflict at one level of magnification is harmony at a higher level. Could it possibly be, therefore, that we—with all our problems, conflicts, neuroses, sicknesses, political outrages, wars, tortures and everything that goes on in human life—are in a state of conflict which can be seen in a larger perspective as a situation of harmony?

Well, it is claimed, you see, that some human beings have broken through to that vision. That they slipped, somehow or other, into states of consciousness where they see the apparent disintegration and disorganization of everyday life as the functioning of a totality which, at its level, is completely harmonious. And you could say, "Aha! At last, I see. I got the point. I've seen how all this makes sense." But what this insight depended upon was your overcoming the illusion that space separates things.

We tend to see the universe itself as really consisting of all the stars and galaxies. That's what it is. That's what we notice. But the space in which all this happens is sort of written off as something that isn't really there. But what one has to realize is that the space is an essential function of the things in the space. After all, you can't have separate stars unless there is a space around them. Eliminate the space and you would see you couldn't have this phenomenon at all—and vice versa. You couldn't have the space—it wouldn't be there in any sense whatsoever—if there weren't the bodies in it. So the bodies in the space and the space are two aspects of a single continuum. They're related to each other in exactly the same way as a back and a front, and you just don't get one without the other.

So the moment you see that intervals—that space—is connective, you can understand at once how you are not just to be exclusively defined as a flash of consciousness that occurs between two eternal darknesses. This is the popular common-sense view that Western people have of their own lives: that you consider that in the darkness that comes before your birth there was no you, and in the eternal darkness that follows your death there is, likewise, no you. And I'm going to discuss these matters not by appealing to any special, spooky knowledge—as if I had been traveling on the higher planes and knew all my previous incarnations and therefore could tell you authoritatively that you are much more than this individuality. I'm going to do it on the basis of complete common sense that everybody has access to the facts and that just what you have to realize is that life is a pattern of immense complexity and what you call "yourself," as a living organism—say, I am my whole body, at the very least—now what is that body? That body is recognizable, and I recognize my friends when I meet them again (with luck), and you recognize me. Although, the last time any of you saw me, I was absolutely something entirely different from what I am now—just as the flame of a candle is never a constant. A flame of a candle is a stream of hot gas. Only, you say "the flame of the candle" as if it were a constant. Well, it is a recognizably constant pattern: the spear-shaped outline of the flame and its coloration is a constant pattern. But in exactly the same way, we are all constant patterns, and that's all we are; the only thing constant about us at all is the doing rather than the being. It's the way we behave, the way we dance. Only, there's no "we" that dances—there's just the dancing. Just as the flame is the streaming of hot gas, just as a whirlpool in a river is a whirling of streaming water. There is no thing that whirlpools; there is simply the whirlpool.

And in the same way, each one of us is a very, very delightfully complex undulation of the energy of the whole universe. Only, by process of miseducation we've been deprived of the knowledge of that fact—not as if there was someone to blame for this, because it's always with our own tacit consent. Because life is, basically, a game of hide-and-seek. Because life is pulsation: on and off, here it is and now it isn't. And by being this pulsation, we know it's there. See, you don't know what you mean by "on" unless you know what you mean by "off." That's why, when we want to awaken someone, we knock at the door. It's not enough to slam the door once with your fist and make this big noise, but you keep up a pulsation. Because that, by its on-and-offness, attracts attention.

All life, you see, is this flickering in and out. Only, there are enormous rhythms in it. There are very fast, flickering ins and outs like the reaction of light upon our eyes, such that when I take a lighted cigarette in the dark and I spin it, you will see a circle of fire. Because the reflection of that cigarette tip on your retina lasts; it endures.

So in this way, very fast impulses are looked upon as constant. And we see—where there are fast impulses—a solid thing. When you look at the blade of a propeller or an electric fan, the separated four or three blades become a solid disk and you cannot throw an egg through it. Well, so in exactly the same way, you can't put your finger through a rock because the rock is moving too fast for your finger to go through. That's the meaning of the whole phenomenon of hardness. Hardness in nature is immense energy, but it acts in a very concentrated space, a restricted space. That's why you can't get through it.

Now, from those very tiny, fast rhythms, which give us the impression of continuity, there are also—in this universe—immensely slow rhythms, and these are difficult to keep track of. And they impress us and depress us as our own life and death, as our coming and going which goes for what is—to us—such a slow pace that we can't possibly believe that it is really a rhythm. We think of it as our birth, as something quite unique that could never occur again, because we're so close to it, you see? And it's moving so slowly. And so, with that point of view, we are like Marshall McLuhan has said—he borrowed a metaphor from me—which is that we are driving a car looking at the rearview mirror. That means that the environment in which you believe yourself to exist is always a past one: it isn't the one you're actually in. The process of growth, the basic process of biology, is one in which lower orders are always being superseded by higher orders. But the lower order can never figure out—or only rarely figure out—what the higher order is that's taking over and may see it as a terrible threat, as total disaster, as the very end. But it can never be aware that the principle of growth always has, and always will, continue. Because that's what's going on. But you never know what the next step is going to be, because if you did know, you wouldn't take it—because it would already be past. Do you understand this? That any certainly known future is an event of which we can say you've had it, and in that sense, it's past.

When we play games—say, in chess, in bridge, or whatever game you're playing—the outcome of the game becomes certain; we at that point cancel the game and begin a new one. Because the whole zest of the thing—and which takes me back to the idea that this whole thing is a hide-and-seek game—is that you don't know what the next order coming up is. But one thing you can be sure of: it will be an order, and it will comprehend you.

At the moment, we stand at a time in history when we're beginning to think of the great countdown to the end of the human race. This is a terrifying possibility that, through atomic energy, we may obliterate this planet and turn the whole globe into a star. Maybe that's the way all the stars started. Imagine, you know, this great thing coming up and the countdown to the end: seven, six, five, four, three, two, one, PEEEEERRRRRRRRRUUMMMMM! Ssshhhhhhhhhwshwshwshwshwwww… POOOSSSSSHHHHH! Ssshhhhwwwwwrrrhhh… POOSSSHHHH! Where have you heard that before? You sit on the seashore, and you hear the waves going in and out. And you don't stop to think. That's what you're doing. That's what the whole business is doing. And there are places where the wave mounts and mounts, and it gets too big for its boots or whatever, and it spills and breaks. We could do just that. But… it's very important to realize that that's what you're doing, because then you don't get panicky about it. And the person who's going to press that button is the person who's going to be in panic.

So if you realize that that's what it is, and that it doesn't really matter if the whole human race blows itself up, then there's a chance that it won't do it. That's the only chance we have. Not to do this thing, which attracts us like a kind of vertigo, like a person who looks over a precipice and is all set to throw himself over, or a person who jumps out of a plane when they're skydiving and forgets to pull the parachute ring because he gets fascinated with a target. It's called target fascination; you just go straight at it, you see? So we can get absolutely fascinated with disaster, with doom. All—you know—all the news in the newspapers is invariably bad news. There is no good news in the newspaper. People wouldn't buy a newspaper consisting of good news. And the fascination, you see, for this doom might be neutralized if we would say, "Well, why bother about that?" It's just another fluctuation in this huge, marvelous, endless chain of our own selves and our own energy going on.

See, here's the problem: because of our myopia, because of the way we've restricted consciousness to focus upon just that certain little area of experience that we call "voluntary action"—that's us—and everything else happens to us. Now, that's obviously absurd. Let's suppose you take in your hand one of those toys—a gyroscopic top—and you suddenly notice, the minute you get this in your hand, it has a kind of vitality to it. It seems to resist you. It starts pushing you in a certain way, see? And sometimes, you're with it and following it, and then sometimes—you see, it's just as if you held a living animal in your hand. You know, you pick up a hamster or a guinea pig, and you hold this little thing in your hand and it's always trying to escape. So the gyroscope always seems to be trying to escape your hold. Now, in exactly the same way, what you're experiencing all the time is that all sorts of things are getting out of control and doing things you don't expect. It's trying to escape your hold. All right, then don't grab it so hard! And you discover that this living thing that you're feeling—like the gyroscope top—it's your own life. Because you can see very simply that you would not understand the experience that you call "voluntary action" and "decision," "being in control" and "being yourself," unless, in opposition to that, there were something else. You couldn't realize self and control and will unless there were something other, out of control, and instead of will, won't! It's the two, together only, that produces the sensation that you call "having a personal identity."

Only, there is a funny thing about human consciousness that has been worked out very carefully in Gestalt psychology, which is that our attention is captured by the figure rather than the background, by the relatively enclosed area rather than the diffuse area, and by something moving rather than what is relatively still. And to all those phenomena that, in this way, attract our attention, we attribute a higher degree of reality than the ones we don't notice. That's only because, for the moment, those are more important to us. Consciousness, you see, is a radar that is scanning the environment to look out for trouble just in the same way as a ship's radar is looking for rocks or other ships. And the radar, therefore, does not notice the vast areas of space where there are no rocks, no other ships. So in the same way, our eyes—or rather, the selective consciousness behind the eyes—only pays attention to what we think is important.

I am, at this moment, aware of all of you in this room, of every single detail of your clothing, of your faces and so on, but I'm not noticing it all. And therefore I will not be able to remember tomorrow exactly what each one of you looked like and what you were wearing. Because what I notice is restricted to things that I think are particularly important. If I notice some particularly beautiful girl in the audience, then I might notice also what she's wearing, and that would be memorable. But by and large—you see—we scan things over, but we pay attention only to what our set of values tells us we ought to pay attention to. And so in this way, we have this rather myopic way of looking at things and we screen out from attention anything that is not immediately important to a scanning system based on sensing danger. But, quite obviously, you—as a complete individual—are much more than this scanning system. You are in relationships with the external world that, on the whole, are incredibly harmonious.

Going back to this illustration of every living body as something like the flame of a candle: the energies of life—in the form of temperature, light, air, and food, and so on—are streaming through you all at this moment in the most magnificently harmonious way. And you're—all of you—far more beautiful than any candle flame. Just sitting in these chairs, just zzzhwwwwt, just going, you know? Only, we're so used to it, we say, "So what? Show me something interesting. Show me something new." Because it's a characteristic of consciousness that it ignores stimuli that are constant. And therefore we eliminate systematically from our awareness all the gorgeous things that are going on all the time and instead only become focused on the troublesome things that might happen to upset it. Which is all right, but we make too much of it and become... we make so much of it that we identify our very selves—I, ego—with the radar, with the troubleshooter. And that's only a tiny fragment of one's total being.

So if you do become aware that you are not simply that scanning mechanism, but you are your complete organism, then—very swiftly in turn, as a consequence

Cyclical

of that—you become aware that your organism is not the way you think about it when you look at it from the standpoint of conscious attention, from the standpoint of the ego. From the standpoint of the ego, your organism is your—kind of—vehicle, your automobile, in which you go around. But from a physical point of view, your organism is, again, like the candle flame or the whirlpool: it is something that is a continuous patterning—or activity—of the whole cosmos.

The key idea here is pattern. Let's suppose—I'm going to borrow a metaphor from Buckminster Fuller—we have a rope, and one section of this rope is made of Manila hemp, the next section is cotton, the next section is silk, the next section is nylon, and so on. Now we tie a knot in this rope—just an ordinary one-over knot—and you find, by putting your finger in the knot, you can move it all the way down the rope. Now as this knot travels, it's first of all made of Manila hemp, it's then made of cotton, it's then made of silk, it's then made of nylon, and so on. But the knot keeps going on. That's the integrity of pattern—the continuing pattern, which is what you are. Because you might, you know, be—for several years—you might be a vegetarian, and you might be a meat eater, and so on. And, you know, your constitution changes all the time, but your friends still recognize you, because you're still putting on the same show. It's the same pattern that is the recognizable individual.

But we are trained in our language. The very structure of the language we talk deceives us into misunderstanding this, because when we see a pattern, we ask, "What's it made of?" Like, you see a table: is it made of wood or is it made of aluminum? But then, when you inquire into what is wood and how wood differs from aluminum, the only thing a scientist can tell you is the different patterns—that is to say, the different molecular structure of the two things. And the molecular structure is not a description of what something is made of, it is a description of what dance it is performing, what motions, what kind of a symphony this is. Because, basically, all phenomena of life are musical, and gold differs from lead in exactly the same way that a waltz differs from a mazurka: it's a different dance.

That is a deception we get into because we have two parts of speech in our grammar: we have nouns and verbs. And verbs are supposed to describe the activities of nouns. And this is simply a convention of speech. You could have a language with only verbs in it; you don't need any nouns. Or you could also have a language with nouns only and no verbs, and it would perfectly adequately describe what's going on in the world. So if you were used to speaking with a language that had one part of speech, you could say just as much as we can with two and be a lot clearer—only, at first it would sound awkward, but you'd soon get used to it. And then, when you got used to it, it would be a matter of common sense that the patterning of the world is not some kind of stuff that's patterning; you don't have to seek a substance underlying the whole thing. It's just patterning! And we're all that.

And so, in this way, there is—to a person who really wakes up—you very soon realize that your existence is not something that is just the hopeless little creature that's suddenly confronted with a great big external world that goes GAAAH! at it—you know?—and eats him up. Every tiniest little thing that comes into being—every minute little fruit fly or gnat or bacterium—I will go so far as to say is an event upon which this whole cosmos depends. Because this thing goes both ways: it's not only that every little organism which exists depends on its total environment. The reverse is also true: that the total environment depends on each and every one of those little organisms. So that you could say this universe consists of an arrangement of pattern in which every event is essential to the whole thing.

Now, we screen that idea out of our consciousness in exactly the same way that we screen out the perception of space as an important reality. Just as we pay attention to the figure and ignore the background, so we see one way of looking at things: mainly, that the organism is very frail against the environment. It lasts a long time—the environment—but the organism only lasts a short time. What do you mean, the environment lasts a long time? What does the environment consist of? Just a lot of little things. And yet, there is the environment just the same way that there is the face in the newspaper photograph behind all those little dots. When you get far enough away from it, you see the face. When you get far enough away from all the organisms and the little bits of things, you see the environment in another scale of magnification. But actually, the whole thing is arranged in a polar system where the enormous depends on the tiny and the tiny depends on the enormous, and you get a relationship between these extremes that can be called a transaction.

So you always—wherever you are looking at the general panorama of sensory experience—try switching. Try shifting your attention to all the things you thought were unimportant—to the constants, to the background—and begin looking at the spaces between people. All painters have to learn this, because you actually have to paint in the background. Weavers know this, because when they are making patterns in weaving, they've got to weave the background as well. Or if you do needlepoint with embroidery, think of the hours you spend putting in the background over the canvas in wool. And you become aware of it. They're much more aware of the background as constituting an essential part of the total experience.

So as you become aware of this, you see the same thing that you notice in music—namely that it is only as a result of hearing the interval between tones that you hear any melody. If you don't hear the interval, you're tone-deaf, and all notes are the same noise; all you hear is rhythm if you don't hear any melody. You've got to hear the interval. So then, watch the intervals between people, the things that aren't said, the things that are tacit, the things that are implicit rather than explicit in all life. And then you begin to get connected. And this is the way it fundamentally comes out of seeing the thing you forgot.

You know, you can always bug people in a beautiful way—in a helpful way—by just saying to them, "What did you forget?" They say, "Well, I don't know. Was I supposed to remember?" "I'm really not trying to put you on. I mean, it's not difficult; this is something completely obvious that you forgot. You'll easily remember it because it's so obvious." Well, that's the hardest thing in the world to think of. What's the most obvious thing I forgot? Huh, what's that? Well, who do you think you are? Well, how do you answer that question, "Who are you?" Well, you give a name. You say, "I'm Joe Dokes. I'm Alan Watts." That's not true. That's what people told you that you were. They put that name on you, and they taught you to identify with it and to behave as it was expected to behave. But that's not who you are. You know very well. Go back in your memory, go back into your infancy before they started telling you all this stuff. Who are you? And if you get with that, you'll know very well who you are: the jolly old ancient of days.

Now, the thing is, we allowed one person, you see—one human individual—to be the incarnate God, because we have all been living in a theory of the universe in which the individual is simply involved in something that happens to them. And we feel that this thing that happens to us is reality; it is facts that we have to face and accept and cope with. It's always something other than you. You don't recognize it as an integral part of your own being without which you cannot know what you mean by the word "I." But the truth of the matter is, that if you face it, every single one of us knows that that isn't true. There is, as it were, a recess of the soul—of the psyche—where everybody knows perfectly well that you are not just this irresponsible little mouse that's been chucked down into this world, but that you are really doing this work. You're running it.

Only you can't admit it just in the same way as you can't admit that you're responsible for the way your own heart beats. You say, "Oh that's not my doing. I've no control over my heart." Do you have any control over being conscious? When you say, "I intend to take my hand down from my face and put it on my leg"—I can do that, but I don't know how the hell it's done. So that's what we mean by the capacity of voluntary control—in the ordinary sense of the word—we don't understand it at all! So you might say, in a funny backward way, that the only kind of control you really understand is when you're not using your will, because you just do it. So easy, like you open and close your hand. You know how to do it? Sure, you know how to do it. But you can't put it into words and explain to someone how to do it. You say, "Well, come on. Aren't you human? Don't you know how to open and close your hand? Just do it, silly!"

But we don't realize, you see, that just as we know how to do this, we know equally well how to turn the Sun into light, how to blue the sky, how to blow the wind, how to wave the ocean, how to digest food. And, I might add, to be digested—by bacteria—and transformed. But the pattern keeps going. And it's always you. Only, you see, you have this marvelous capacity to transform yourself without knowing that you're doing it. Therefore, you keep surprising yourself, and therefore you keep on doing it. Because if you didn't surprise yourself, you wouldn't go on doing it. It's just the very fact, you see, that you seem to be the victims of a thing you don't understand and that you seem to conclude your life every time in a wipeout called "death"—where all your control goes—it's just exactly that opposite condition to what you call "being alive" that allows you to be alive! Only, every time it happens, it's like it's new. It's like every time you're born, it seems like it was the only time. But of course, if it wasn't like that, you wouldn't do it.

Contributors

Editors

Daniel Kwan is a writer, director, and dad known for co-writing and co-directing movies under the moniker DANIELS. He's overwhelmingly good at dancing which is a bit of a burden he has to carry.

Daniel Scheinert is a writer and director, but wanted to be an actor as a kid, so now he just makes cameos in his own films, like playing Dick in *The Death of Dick Long* and Paul Dano's uncredited body double in *Swiss Army Man*.

Meg Miller is a writer and editor based in Berlin. She has contributed to *The New York Times, The Atlantic*, The Creative Independent, *Bulletins of the Serving Library*, and other publications and books, mostly about the ways design, art, language, and technology shape culture and society. She is editorial director at Are.na and a contributing editor for Source Type.

Writers

Jorge Luis Borges was born in Buenos Aires in 1989 and published many collections of poems, essays, and short stories before his death in Geneva in June 1986. He was director of the Argentine National Library from 1955 until 1973.

Billy Chew is a screenwriter, secret songsmith, and ginger cookie cook known for writing *The Death of Dick Long* and a dizzying number of beloved unproduced screenplays with similarly unmarketable titles.

David Eagleman is a neuroscientist, *New York Times* best-selling author, TED speaker, and Guggenheim Fellow. He is the writer and presenter of the Emmy-nominated series *The Brain* on PBS, as well as *The Creative Brain* on Netflix. In Palo Alto, California, he teaches at Stanford University, runs a startup neurotech company called Neosensory, and directs the Center for Science and Law. He is the author of eight books, including the international bestsellers *Sum* and *Incognito*, and his newest book, *Livewired*.

Kelsey Keith is a writer and editor based in Berkeley, California. She's currently the editorial director at Herman Miller, following a five-year run as *Curbed*'s editor-in-chief. She's a contributing editor at *Elle Decor* and the executive producer of *Nice Try!*, a podcast that explores our attempts at creating utopia.

Born in Tel Aviv in 1967, **Etgar Keret** is a leading voice in Israeli literature and cinema. A *chevalier de l'ordre des Arts et des Lettres*, and recipient of the Charles Bronfman Prize, the Sapir Prize and the Caméra d'Or at the Cannes Film Festival, he is the author of the memoir *The Seven Good Years* and of six internationally bestselling story collections, including, most recently, *Fly Already*. His work has been published in more than 45 languages. *The Middleman*, a French speaking mini-series co-written and co-directed with Shira Geffen, appeared to acclaim in May 2020.

Julia Pott is a writer, illustrator, animator, and fact collector from England known for creating the show *Summer Camp Island* (and voicing the head witch of the Summer Camp), and illustrating unforgettably sad animals.

Sasha Sagan is a writer and speaker based in Boston, Massachusetts. Her essays and interviews on death, history, and ritual through a secular lens have appeared in *The Cut, O, the Oprah Magazine, Parents, Atmos,* and more. She is the author of *For Small Creatures Such As We: Rituals for Finding Meaning in Our Unlikely World*. You can follow her on Twitter and Instagram @SashaSagan.

Emily Segal is an artist, writer, and trend forecaster based in Los Angeles. Her debut novel, *Mercury Retrograde*, a *New York Times* Editor's Choice, was published in 2020 by Deluge Books. She co-founded the trend forecasting group K-HOLE, and currently co-directs the consultancy, Nemesis. Segal's novel-in-progress, *Burn Alpha*, was the first book to be crowdfunded with cryptocurrency, under the token $NOVEL, in 2021.

Lizzy Stewart is an author and illustrator living in London. She has written and illustrated a number of picture books for children, and her debut full-length work for adults, *Alison,* was published in 2022.

Esmé Weijun Wang is the New York Times-bestselling author of *The Collected Schizophrenias: Essays and The Border of Paradise: A Novel.* She received the Whiting Award for Nonfiction in 2018, was named one of Granta's Best of Young American Novelists of 2017, and won the Graywolf Nonfiction Prize in 2016. Esmé can be found at esmewang.com, where she also teaches writing online.

Alan Watts was a prolific author and speaker, and one of the first to interpret Eastern wisdom for a Western audience. Born outside London in 1915, he became an Episcopal priest for a time, and then relocated to Millbrook, New York, where he wrote his pivotal book *The Wisdom of Insecurity: A Message for an Age of Anxiety*. In 1951 he moved to San Francisco where he began teaching Buddhist studies, and in 1956 began his popular radio show, "Way Beyond the West." By the early '60s, Alan's radio talks aired nationally and the counterculture movement adopted him as a spiritual spokesperson.

Natalie Wolchover is a senior writer and editor at *Quanta Magazine*, with bylines in *Nature, NewYorker.com, Popular Science,* and other publications. Her writing has been featured in *The Best American Science and Nature Writing* and *The Best Writing on Mathematics,* and has won several awards. Hailing from both London, England and rural Texas, she now lives with her wife, daughter, and two cats in Queens, New York. She wrote the introductions to each of the sections, in addition to serving as the book's science consultant.

Illustrators

Robert Beatty is a Lexington, Kentucky-based artist and musician known primarily for his prolific work in the field of contemporary album cover artwork. He has designed well over 100 record covers for the likes of Tame Impala, Kesha, The Flaming Lips, Oneohtrix Point Never, and countless others. Beatty's illustration work has been featured in publications as diverse as *Lucky Peach, Surfer Magazine, Wired, The Oxford American, The New Republic,* and *The New York Times.*

Jun Cen is a New York-based award-winning illustrator. He received a gold medal from the Society of Illustrators and a bronze from the ADC awards. His work can be found in *The New York Times, The New Yorker, The Washington Post,* and more.

Liam Cobb studied illustration at Camberwell Art College, UAL. He has published several comics and works as an illustrator for clients such as *Dezeen, The New York Times, Wired,* and WeTransfer. He is also a background designer on animations, including *The Midnight Gospel*.

Céline Ducrot studied graphic design in Biel, Switzerland and at the Academy of Fine Arts Leipzig, Germany, where she finished her diploma in illustration in 2017. She has won numerous awards for her work, including the Swiss Design Award in 2018, the Anderfuhren Grant 2018, and a DAAD Award for exceptional foreign students in 2015. Since 2017, she has been working as an independent illustrator and artist in Switzerland. More of her work can be found at www.celineducrot.ch and on Instagram @celine.i.d.

Derek Ercolano is a designer and illustrator originally from New York, currently located in Oslo, Norway. He can be found on Instagram @FragileMagic.

Max Guther is a Berlin-based illustrator and designer using a unique 3D aesthetic to bring his slightly hyperreal illustrations to life. Influenced by the aesthetic of old computer games and obsessed with the isometric perspective, he creates worlds with the focus on both interior and exterior design, but also on the minute details of everyday life, capturing a larger truth about human behavior. Max can be found at maxguther.de and on Instagram @maxguther.

Ram Han is a digital painter based in Seoul. She is interested in how experimental fantasy can be injected into pop culture, subculture, and media. She believes the nature of memory is the ambiguity between virtual and real life, and her narrative goal involves owning the memory of a place or experience that a viewer never had. Ram can be found at ram-han.com and on Instagram @ram__han.

Ilya Milstein is a self-trained Australian illustrator working in New York. Employing a combination of traditional and digital processes in a style reminiscent of Franco-Belgian comics and Japanese woodblock prints, his drawings are often highly-detailed, dense with arcane references, and nostalgic in their character. His work has been recognized by the Society of Illustrators, American Illustration, Communication Arts, and The One Club for Creativity; and he has created work for *The New Yorker, The New York Times,* Apple, and The Metropolitan Museum of Art.

Jordan Moss is a Brooklyn-based illustrator and graphic designer. With a background in fine art and advertising, she loves getting the chance to design in all forms and mediums. Whether it's hand-painted murals in her hometown or digital stage graphics at Coachella, she's excited to create in all spaces. You can stay up to date with her at jordanemoss.com

Jul Quanouai is a French artist and illustrator who creates drawings and paintings focused on objects, nature, movements, and sensations. As an illustrator, Jul works with *Zeit Magazin, Brick Magazine, The Baffler, Kapsel Magazin, Revue Lagon,* Pli Éditions, and Drag City Records. Jul's work has been published and shown by RFI gallery, Galerie M, Colorama, Super-structure, and Éditions Matière. You can find Jul on Instagram @julquanouai.

Qieer Wang is a Chinese artist exploring abstract visual expressions that create emotional connections. She is the founder of QIE MEDIA, a creative animation studio in TV and advertising, and is a visual development artist working on animation features for Netflix.

Finally, at the end of all our wanderings, we return to our tiny fragile, blue-white world, lost in a cosmic ocean vast beyond our most courageous imaginings. It is a world among an immensity of others. It may be significant only to us. Carl Sagan, Cosmos

A Vast, Pointless Gyration of
Radioactive Rocks and Gas
in Which You Happen to Occur

Edited by Daniel Kwan and Daniel Scheinert

A24 Films LLC
31 West 27th Street
New York, NY
a24films.com

Head of Publishing
Perrin Drumm

Editor
Meg Miller

Art Director
Kyra Goldstein

Science Consultant
Natalie Wolchover

Poetry Consultant
Verônika Shülman

Copy Editor
Danielle Carter

Proofreader
Todd Barringer Albright

Production Management
Actualizers

Special Thanks
Carl Sagan
Puddle the dog

Book Design
Chris Svensson
chrissvensson.info

Type
Söhne, Söhne Schmal, Söhne Breit,
and Söhne Mono by Klim Type Foundry

Paper
150 gsm Condat Matt Périgord
115 gsm Munken Print White 15
60 gsm IBO

Printer
die Keure, Belgium
diekeure.be

Illustration
Robert Beatty
Jun Cen
Liam Cobb
Derek Ercolano
Céline Ducrot
Max Guther
Ram Han
Ilya Milstein
Jordan Moss
Jul Quanouai
Lizzy Stewart
Qieer Wang

Licensed Materials

Kaveh Akbar, "Portrait of the Alcoholic Stranded Alone on a Desert Island" from *Calling a Wolf a Wolf*. Copyright © 2017 by Kaveh Akbar. Reprinted with the permission of The Permissions Company, LLC on behalf of Alice James Books, www.alicejames.org.

Jorge Luis Borges, "The Book of Sand," copyright © 1998 by Maria Kodama; translation copyright © 1998 by Penguin Random House LLC; from *Collected Fictions: Volume* 3 by Jorge Luis Borges, translated by Andrew Hurley. Used by permission of Viking Books, an imprint of Penguin Publishing Group, a division of Penguin Random House LLC. All rights reserved.

Gwendolyn Brooks, "Speech to the Young: Speech to the Progress Toward." Reprinted by Consent of Brooks Permissions.

Ursula Le Guin, "Hymn to Time" Copyright © 2016 by Ursula K. Le Guin. First appeared in *Late in the Day*, published by PM Press in 2016. Reprinted by permission of Ginger Clark Literary, LLC.

Etgar Keret, "Broken" copyright © 2022. All rights reserved

Maggie Smith, "Good Bones" from *Good Bones: Poems*. Copyright © 2017 by Maggie Smith. Reprinted with the permission of The Permissions Company, LLC, on behalf of Tupelo Press, Tupelopress.org.

Raymond Queneau, "Cent mille milliards de poèmes" © Editions Gallimard 1961. English translation from *The Penguin Book of Oulipo*, edited by Philip Terry, published by Penguin Classics. Copyright © 2019. Reprinted by permission of Penguin Books Limited.

Alan Watts, "We As Organism" from the *Philosophy and Society* lecture series at alanwatts.org. (The title of this book is also taken from this lecture.)

Night sky photograph used in background of chapter title pages courtesy of Lubo Minar / Unsplash.

Screenshots used in "Who Made the Bubble Maker?" booklet from *Cosmos: A Personal Journey* Season 1, Episode 10 "The Edge of Forever," PBS, 1980.

© 2022 the authors, editors, and owners of all respective content.

All rights reserved; no part of this publication may be reproduced, stored in a retrieval system, or transmitted in any form or by any means, electronic, mechanical, photocopying, recording, or otherwise, without prior written consent of the publisher.

Every effort has been made to identify copyright holders and obtain their permission for the use of copyrighted material. The publisher apologizes for any errors or omissions and would be grateful if notified at info@a24films.com of any corrections that should be incorporated in future reprints or editions of this book.

ISBN 978-1-7359117-7-9